Inorganic Trace Analysis:

PHILOSOPHY AND PRACTICE

Inorganic Trace Analysis:

PHILOSOPHY AND PRACTICE

A. G. HOWARD and P. J. STATHAM
University of Southampton
Southampton
Hampshire
SO9 5NH
UK

JOHN WILEY & SONS
Chichester • New York • Brisbane • Toronto • Singapore

Other Wiley Editorial Offices

John Wiley & Sons, Inc., 605 Third Avenue,
New York, NY 10158-0012, USA

Jacaranda Wiley Ltd, 33 Park Road, Milton,
Queensland 4064, Australia

John Wiley & Sons (Canada) Ltd, 22 Worcester Road,
Rexdale, Ontario M9W 1L1, Canada

John Wiley & Sons (SEA) Pte Ltd, 37 Jalan Pemimpin #05-04,
Block B, Union Industrial Building, Singapore 2057

Library of Congress Cataloging-in-Publication Data

Howard. A. G. (Alan G.)
Inorganic trace analysis : philosophy and practice / A. G. Howard and P. J. Statham.
p. cm.
Includes bibliographical references and index.
ISBN 0 471 94144 1
1. Trace analysis. 2. Inorganic compounds—Analysis
I Statham, P. J. II. Title.
QD139.T7H69 1993
543—dc20 93–24651
CIP

British Library Cataloguing in Publication Data

A catalogue record for this book is available from the British Library

ISBN 0 471 94144 1

Typeset in 10/12pt Times from author's disks by Text Processing Department, John Wiley & Sons Ltd, Chichester
Printed and bound in Great Britain by Biddles Ltd, Guildford, Surrey

Contents

Preface

As our understanding of the physics, chemistry and biology underlying natural and man-made objects and phenomena improves, there is a growing awareness of the significance of the presence of elements which occur at very low concentrations. Trace measurements are therefore required in a wide range of disciplines, from medical diagnosis, through veterinary science and the semiconductor industry, to the environment. Concern over the build-up of potentially toxic materials in the food chain, for example, has led to improved environmental quality standards and an increasingly challenging task for the analyst. There are many cases in which the conventional view of how a system works has been overturned as a result of data generated by newly developed analytical techniques. This in turn has led to a demand for more accurate data to be generated at lower concentrations.

The skills, effort and understanding required for methods which are employed in an analysis are not in themselves directly related to concentration, but as the concentration of the analyte decreases additional difficulties rapidly become evident. At these levels even apparently small factors in the design of the analytical method can have a major effect on the analysis. A speck of dust may significantly contaminate the sample, or adsorption onto a container wall may result in total loss of the analyte. The development of clean sampling, handling and analytical procedures, which in themselves can involve a substantial investment of time and resources, has been a necessary step towards obtaining accurate measurements in a number of fields. The scientific journal is not normally considered to be the place for the publication of those features of an analysis which can be classified as skill, let alone the place to discuss the general philosophy and guiding principles behind trace analysis. This assumed knowledge can, however, be critical to the quality of data which are generated.

This volume concentrates on inorganic trace analysis, with an emphasis on the analysis of metals. It does *not* set out to provide a detailed set of analytical procedures, which may rapidly become obsolete, nor does it provide specific details of particular techniques which are used at the determination stage, as these are widely covered in specialized books on instrumental analysis. Its objective is to help develop a thinking approach to trace analysis by pulling together scattered information on the techniques and materials which are used in trace analysis and by identifying the underlying principles behind the development of trace analysis procedures and their use. In this way the book provides a grounding for those new to the trace level aspects of inorganic analysis, and gives a source of additional ideas and information for workers already in the field. With the accent on the underlying principles and the philosophy behind method development, rather than a series of recipes, the intention is to provide a book that will not date rapidly but will stimulate others to push forward the frontiers of trace analysis.

A basic level of knowledge of analytical and general chemistry is assumed. Although primarily directed towards the analyst, the philosophies, handling techniques, purification methods and information on materials which are contained in this book will hopefully also be of use to workers in other disciplines which are susceptible to contamination.

Many of the ideas which are presented in this book have developed through the years as a result of stimulating discussions with colleagues in the analytical community. We are particularly indebted to Dr Peter J. Ovenden for his astute comments and accurate proof- reading of the manuscript during its preparation.

This book is dedicated to Liz, Clare and Catriona Statham for their forbearance while the wordprocessor dominated family life.

1 Introduction

Trace analysis both leads and follows the advances of science, each new analytical development bringing with it additional achievements and expectations for the future. Radical developments occur only rarely in all areas of science and it normally takes time for even the smallest advance to permeate through to the routine laboratory. The levels of analyte which can be determined routinely have dropped significantly in recent years, and what was considered impossible twenty years ago has now become commonplace. This progression has been brought about, rather surprisingly to some, as much by a gradual development of sample handling techniques as by radical leaps in technology. General improvements in sample handling procedures and the performance of electronic and optical systems have contributed equally to a progressive improvement in the performance of the instrumental aspects of trace analysis. Whether driven by, or arising from, low level analytical measurements, these enhancements have resulted in an increased demand for high precision analyses of complex matrices. Without such measurements our understanding of environmental chemistry, trace element metabolism and enzyme activity or the construction of semiconductors would have been severely curtailed.

The significance and origins of the advances which have made trace analysis possible are often forgotten. Many minor developments have occurred which are now taken for granted as standard practice, and the advances made in one area of expertise have fed into other areas. The development of modern day electronics is dependent upon the availability of extremely high quality materials, and to ensure the purity of these materials the analyst in turn must have available reagents and procedures which are compatible with the stringent requirements of the analysis. This will normally mean that the analyst must have available material which is of a higher purity than the samples to be analysed.

Our understanding of the chemical composition of the environment has changed radically with the development of analytical techniques. Over the past decade the concentrations of many trace constituents in the deep ocean have appeared to drop by orders of magnitude (Figure 1.1). Needless to say this has little to do with a real drop in concentration, but reflects improvements in sample handling and analytical technique.

As a body of accurate data has accumulated, general trends have been revealed in the distribution of trace elements through the oceanic water column which could not have been deduced from poor data sets (Figure 1.2). The recognition of such trends in the data has in turn led to the trend itself being used as a criterion with which to assess the reliability of the data set. If misused, this assumption that a

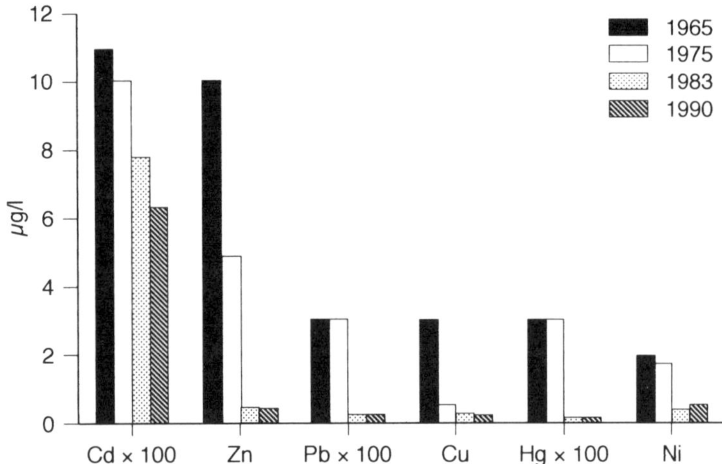

Figure 1.1 The apparently changing levels of dissolved trace metals in deep ocean waters. Data from: Goldberg, 1965; Brewer, 1975; Bruland, 1983; Burton and Statham, 1990

trend should be present could be incorrectly employed as a criterion for the rejection of significant data. Such a problem lead to a delay in the discovery of the depletion of the ozone layer.

Our increasing understanding and awareness of the build-up of potentially toxic trace elements through the food chain, and the importance of sub-lethal effects, has led to legislation designed to restrict exposure to, and consumption of, potentially dangerous materials. In many cases legislated levels have been based on the limitations of current analytical methodology. With the development of improved procedures, however, accessible detection limits eventually drop below background levels and more appropriate guidelines can be issued to limit exposure.

1.1 TRACE AND ULTRA-TRACE ANALYSIS

Trace analysis is under constant development and the impossible analyses of today will become the routine of tomorrow. The description of a procedure as being a trace analysis has more to do with the approaches required to carry it out than with the actual magnitude of the levels involved. Any attempt to define trace analysis in terms of concentration is doomed to failure, and for the purposes of this book the concept of trace analysis will be taken in a wide and flexible sense. We will therefore consider such an analysis to be any analytical procedure which requires special techniques to be employed because of the low levels of analyte which are present. This may involve precautions in handling to minimize contamination, the extensive purification of reagents or the use of instrumentation near to the limits of its performance.

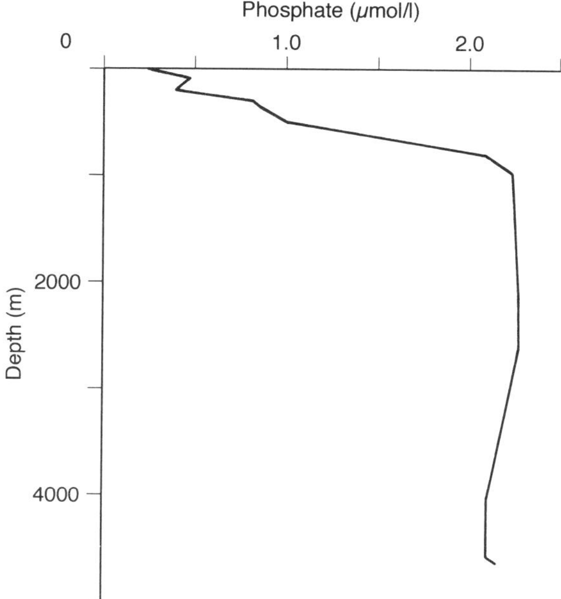

Figure 1.2 Dissolved phosphate in the Indian Ocean water column (Morley *et al.*, 1993))

> TRACE ANALYSIS: An analytical procedure requiring special steps to be taken because of the low concentration of analyte which is present

For a number of elements there has been a demonstrated need to push methods to their absolute limits. To differentiate those analyses which require greater expertise than is required for the more routine low level measurements, these are often described as being 'ultra-trace analyses'. Such analyses will normally require a particularly high level of expertise of the analyst and the highest quality instrumentation. Whilst tomorrow may see the dawning of the 'hyper-trace analysis', it is more likely that the levels associated with the current classifications will reduce to match the levels of expertise and technology which are at that time accepted as normal.

1.2 THE ROLE OF PRECONCENTRATION

Preconcentration is the process by which the concentration of an analyte is increased prior to the determination stage. In the analysis of copper in water, such a preconcentration can be achieved by the extraction of the copper

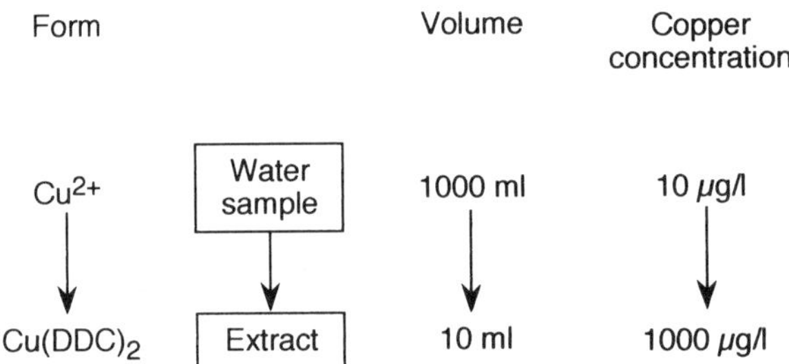

Figure 1.3 The preconcentration of copper as its diethyldithiocarbamate complex

as its diethyldithiocarbamate (DDC) complex into a smaller volume of 4-methyl-2-pentanone (methyl isobutyl ketone) (Figure 1.3).

$$Cu^{2+} + 2\ (C_2H_5)_2NCS_2^- \rightleftharpoons Cu[(C_2H_5)_2NCS_2]_2$$

The transfer of the copper from 1 litre of water to 10 ml of extracting solvent enriches a solution containing 10 g of copper per litre by a factor of one hundred, giving a final concentration of 1 mg per litre. This preconcentration takes the analysis well into the working range of flame atomic absorption spectroscopy. Direct analysis of the sample would not have been possible by this technique.

There are many other techniques which can be employed for preconcentration, including collection on solids such as ion-exchange resins, solvent evaporation and co-precipitation.

The direct analysis of a sample without prior treatment must always remain the ideal, as it limits the potential for contamination resulting from the addition of reagents and the transfer of the analyte from vessel to vessel. Few current instrumental techniques can be used directly at the extremely low levels required for some applications and it would be naive to assume that all laboratories throughout the world which are required to carry out trace analyses are equipped to the highest standards. As most instrumental techniques involve in some way the measurement of concentration, for analyte levels below the available working range some means of increasing the concentration is required—the preconcentration step. Even if, as is the case graphite furnace atomic absorption spectroscopy, the analyte response depends on absolute quantities, the technique may in reality be concentration limited as the maximum volume of analyte which can be introduced into the instrument is itself fixed.

> Preconcentration not only concentrates the analyte but may also reduce interferences

It is also unfortunately rare that techniques are completely interference-free and the separation of interfering species from the analyte is therefore necessary. A well designed preconcentration step may not only concentrate the analyte, but at the same time may offer the potential of removing interfering species.

There are additional advantages to be gained from the use of a preconcentration procedure. These include the concentration of samples to levels which are less subject to contamination, and the improved storage and transportability of samples. It is also not uncommon for the products of a preconcentration procedure to be in a more stable chemical form than they were in the original sample, improving the time that the sample can be kept before it is analysed.

There are unfortunately penalties to be paid for the use of preconcentration techniques. Preconcentration normally involves extra manipulations of the sample and additional reagents; each additional step and reagent takes its toll in sample contamination and/or loss. Great care must therefore be taken in the use of preconcentration techniques, and much of this book involves the nature of hazards associated with such manipulations and how to minimize problems.

1.3 SPECIATION

In the past a large proportion of all inorganic analyses has involved the measurement of the total quantity of an element. This has been dictated, in the case of metals, by the available techniques and a failure to recognize the significance of the chemical form in which the element is held. This has not, however, been the case in the analysis of non-metals. Their study has frequently involved the measurement of a particular species. It is normal, for example, to measure dissolved 'reactive' phosphate and not total phosphorus in drinking water. This choice of determinand is dictated more by the use of colorimetric techniques than by the immediate need to measure just one phosphate species rather than total dissolved phosphorus. The measurement technique yields only information on reactive phosphate, and information on total dissolved phosphorus is much more difficult to obtain with any confidence.

In order to bridge this gulf between methods which give the total amount of an element in a sample and the methods which give full structural information on the chemical form of the element, the concept of 'speciation analysis' has been developed. This is used to describe an analytical procedure which provides information on any aspect of the chemical form of an element.

> SPECIATION ANALYSIS: a measurement which gives qualitative and/or quantitative information on a variable aspect of the chemistry of an element

Although the eventual goal of every speciation analyst must be to identify the

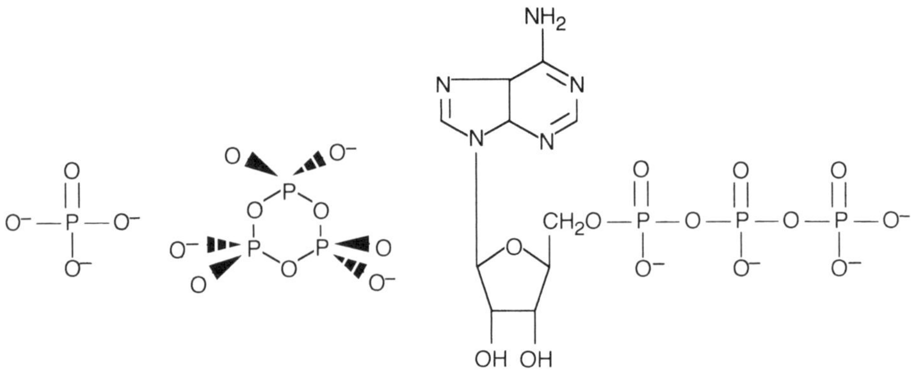

Phosphate Trimetaphosphate Adenosine 5′-triphosphate (ATP)

Figure 1.4 A variety of ions which are both phosphorus phosphate species

full chemical structure of all the chemical forms of the element, any structural information, be it oxidation state, degree of methylation or molecular size, contributes to the appraisal of the speciation of the element.

For phosphorus, typical speciation analyses might measure any one or more of the following: isotopic abundances of the phosphorus, oxidation state, reactive phosphate (anything which reacts to give molybdenum blue in the classic colorimetric analysis), pyrophosphate, organic phosphorus or adenosine triphosphate (ATP) (Figure 1.4).

There are many speciation analyses which are possible for any particular element and the chosen method must be well matched to the requirements of the particular scientific study.

1.4 GENERAL STRATEGY

The quality of data generated in an investigation is only as good as the weakest point in the chain of events from the sampling to the treatment of the final data. The final errors of the analysis are the result of compounded errors which have been picked up throughout the analysis trail. It is therefore important to minimize contributing errors at each step of the analysis.

Analysis problems start with experimental design and sampling. If these two steps are outside the control of the analyst no amount of analytical skill will recover useful results from poorly selected or stored samples.

> The analyst should control or participate in experimental design and be responsible for all sampling.

Detailed consideration of experimental design and sample collection is beyond the scope of this book, but some universal features should be noted.

1.4.1 Planning

Surprisingly often, analyses are requested, or carried out, without any true understanding of the reasons for the study or the features which are to be investigated from the results.

> Unless you know exactly what you want to study you will design a poor experiment.

There are a number of questions which should be answered before proceeding with any practical work.

1. *What is the aim of the study and how are these aims to be met by the proposed analyses?* It may be that the proposed analyses do not provide sufficient information for the results to be adequately interpreted. Supporting measurements may be required. Is information required on the total concentration of lead in soil or is the requirement for an appraisal of the lead which is available to be assimilated by plants? There is a great difference between knowing the concentration of iron in the leaves of a plant and the concentration of iron in the whole plant.

2. *Has the experimental design been considered?* Is it possible that with a better design more information could be extracted from the experiment or that fewer analyses could be performed?

3. *What precision is required?* The greater the required precision the more costly, both in time and resources, the analysis will be.

4. *What accuracy is required?* Many studies are of a comparative nature and accuracy may, under carefully considered conditions, be less of a problem than reproducibility. In other circumstances the reverse may be true.

5. *Has the proposed experiment incorporated sufficient controls*: to assess, for example, contamination or poor analyte collection by a preconcentration technique?

1.4.2 Sampling

To state that the sample should be representative of material to be analysed may at first seem to be stating the obvious. There are many problems to be encountered in trying to achieve this.

> Poorly chosen samples can only generate poor results

Take for example the appraisal of the extent and mechanism of lead uptake resulting from the human consumption of a plant such as a cabbage. In the planning of this experiment it might be decided that the study should be limited to the leaves, as the stalk and roots are not normally consumed by humans. When it comes to the sampling, however, at what age should the plant be sampled and how does one ensure that the plant is sampled in a representative manner? As far as atmospheric fallout is concerned, it is probable that the upper leaves receive greater quantities of the lead, but are they also younger and for that reason less contaminated? Is human consumption limited to the younger leaves?

Having decided on the leaves to be sampled, the next decision is which part of the leaves to analyse. Here the options may be between attempting to measure the quantity of fallout on the surface of the leaf or whether to measure the whole leaf. If this latter course is taken, it should be borne in mind that the results will include lead which has been translocated through the roots to the leaf. It could even be argued that a truer assessment of the danger to human health could be gained only from the measurement of the lead in the cooked leaves, as much of the lead might be removed during cooking.

During the process of sampling it is important that the analyte should be neither lost nor gained. In our example, dust deposits on the leaf may well fall off during collection or may be transferred from leaf to leaf on the hands. Specific precautions must therefore be adopted to prevent the occurrence of such contamination or losses.

> **The uncertainty principle in trace analysis**
> The mere act of taking a sample may change its nature or the concentration of the analyte

Although the most obvious contamination results directly from the reagents and implements employed, there are circumstances in which the mere act of taking the sample may result in a change in concentration. A classic example of this is the taking of blood samples by syringe for magnesium analysis. The problem here is reported to arise by the penetration of the needle into the skin setting off a physiological response which alters the magnesium levels in the vicinity of the wound.

Additional problems may be encountered if the study involves speciation. Changes in speciation can be particularly troublesome in cases where the analyte species is subject to oxidation and in systems which are not at chemical equilibrium. Where an equilibrium between species is present, the mere act of removing the species of interest from solution may result in the re-establishment of the equilibrium by the regeneration of the removed species.

A final point to be made in this general discussion is that there may be only one opportunity to obtain samples.

> Oversample; it is always easier to collect too many samples and throw them away than to repeat the whole experiment

1.5 CONTAMINATION

The most common difficulties which occur in trace analysis are those which involve contamination. The extent of the problem depends on the amount of analyte present in the sample, the procedures which are used, and how ubiquitous the analyte is in the potential sources of contamination.

> Beware; you can add contaminants or they can make their own way into your samples

The worst situations are normally those involving the measurement of very low concentrations of the ubiquitous elements such as iron and zinc. At times it may appear that contaminants are raining down from the sky and into the samples. In reality, this is quite often exactly what is happening!

The magnitude of the contamination is normally assessed by the measurement of 'blanks'. There are several types of blank which can be used but in its most straightforward form the blank experiment can be thought of as involving the analysis of an infinitely small sample. It is of prime importance, as it indicates the level of contamination resulting from the reagents, glassware etc. employed in the analysis. If the blank analyses are within reasonable levels, they can be employed to apply contamination corrections to the results obtained from the analysis of the samples. A more extensive discussion of the application of blank analyses to the diagnosis of contamination problems will be given later.

There are two main contributions to the blank. Firstly, there are the contributions from the reagents and glassware employed in the procedure. These contributions, within a batch of analyses, are normally constant and should not vary from one sample to the other. The blank analysis can therefore be employed as a correction factor to be applied to the other analyses. There is also an element of the blank which has a random nature; this may reflect a particle of dust which fell into the solution or a single faulty container in a batch. Statistical evaluation of the variability of the blank analysis is therefore as important as it is in the analysis of the samples. A single blank determination in a batch of analyses will not be sufficient to provide the statistical information which is required to estimate the error associated with the blank-corrected analytical result.

> Multiple blanks are necessary in all batches of analyses

The effect of a contaminant on an analysis may not be restricted to an increase in the analyte concentration. Contaminating species may not be inactive in the analysis and may interfere in the final determination stage. Alternatively, the presence of only a small quantity of solid material in a solution may provide a very large sorption surface and thus lead to a reduction of the analyte concentration in solution.

The blank is a cumulative factor; every operation on the sample potentially increases its value. It follows therefore that a first step in contamination control is to minimize the number of handling steps. This is also true for the number of reagents; the fewer that are used, the less probable is contamination.

> Reduction of the number of reagents and handling steps reduces contamination

The most critical factor of all in contamination control is the analyst and his or her attitude. Trace analysis cannot be carried out passively. There must be rigorous and careful initial planning and continuous evaluation and self criticism during the work. Every step from a complex solvent extraction to the picking up of a flask must be viewed suspiciously as a potential route for contamination. The working environment, the chemicals and equipment used and the procedures employed, all contribute to problems in trace analysis. Much of the remainder of the book addresses the ways in which this problem can be identified and controlled.

If nothing else:

> **THINK BLANK!**

1.6 LOSSES

Much of the emphasis so far has been placed on the potential contamination of materials which are employed in trace analysis. Contamination problems are derived from the similar levels of analyte which are present in the samples and those which are present in the working environment and the reagents which are employed in the analysis. Losses, on the other hand, are a particularly significant problem in trace analysis. Container surfaces, for example, may present a significantly large area on which the analyte can be adsorbed. At higher levels such a small absolute loss would have little effect on the concentration; at trace levels a large proportion of the analyte may be stripped from solution.

1.7 THE PHILOSOPHY OF TRACE ANALYSIS

The development and validation of a trace analysis procedure involve many stages. It is often believed by those who have received little training in analysis that it is normally possible to take an 'off the shelf' procedure and apply it to a new sample type. This is rarely the case however, and an understanding of the chemistry underlying each stage is therefore invaluable if a procedure is to be developed for the new application. All analytical procedures have at some time been developed from first principles, and it is a valuable exercise to dissect recommended procedures into their component parts and to attempt to identify the reasons for choosing particular stages. In modern analysis every attempt should be made to eliminate those steps which are included for no apparent reason or just because 'it has always been done that way'.

Recognizing that no two analyses are identical, this book sets out to identify the principles that must be employed in the development of new analytical procedures or in carrying out existing ones. It includes a range of factual information which will assist the reader in the choice of method but does not set out to include 'standard methods' which may soon become dated. The assumption is made that the reader has some knowledge of basic analytical chemistry and the instrumental techniques which will be employed in the analyses. It is hoped that the book will prompt the reader to think positively about the steps which have to be put together to develop an analytical procedure. There will be parts of the book which appear facile to some readers. These same sections may however prompt others to an insight into features which have previously not been evident. Overall we hope to give the reader a better understanding of how existing procedures have been constructed, and to develop a positive approach to the production of new procedures. By the development of a questioning and understanding attitude we hope to enhance the quality of trace analysis by removing much of the 'witchcraft' which has crept into many existing methods. At the same time the approach should enhance both productivity and the analysts' involvement and enjoyment in their subject.

REFERENCES

Brewer, P.G. (1975) In J.P. Riley and G. Skirrow (Eds), *Chemical Oceanography*, 2nd Ed., Vol. 1, Academic Press, London.

Bruland, K.W. (1983) In J.P. Riley and R. Chester (Eds), *Chemical Oceanography*, Vol. 8, Academic Press, New York.
Burton, J.D. and Statham, P.J. (1990) In R.W. Furness and P.S. Rainbow (Eds), *Heavy Metals in the Marine Environment*, CRC Press, Boca Raton.
Goldberg, E.D. (1965) In J.P. Riley and G. Skirrow (Eds), *Chemical Oceanography*, Vol. 1, Academic Press, New York.
Morley N.H., Statham P.J. and Burton J.D. (1993) *Deep Sea Res.* **40**, 1043–1062.

2 The Working Environment

2.1 INTRODUCTION

In situ analyses are rarely available in trace analysis, and in most instances the sample must be analysed in a laboratory environment. The focus of this section is on the environment in which these analytical manipulations take place and the control of the component of the analytical blank which arises from these surroundings.

As an appreciation of the contamination problems which arise from the laboratory has grown, a more critical view of the working environment has developed. Problems are primarily associated with contamination transported through the atmosphere, normally in the particulate phase. There is a range of activities, apart from trace analysis, in which atmospheric contamination is a major problem, including the pharmaceutical, precision optical and mechanical engineering and, in particular, the micro-electronics industries. In these activities it is more frequently the fact that viable (viruses, bacteria, spores etc.) and non-viable particles (e.g. corrosion flakes, dead skin cells, minerals etc.) are present, rather than the trace element content of these contaminants, which degrades the quality performance or yield of a product. Procedures which have been developed to control particulate contamination in these applications have been successfully adapted to trace analysis.

This chapter reviews the problem of contamination from the working environment and considers the range of approaches available for its control.

2.2 SOURCES OF CONTAMINATION IN THE WORKING ENVIRONMENT

2.2.1 Airborne Contaminants

The conventional laboratory, particularly if it is situated in an old building, can be a major source of airborne contamination. The contaminants are normally in gaseous or particulate form, but examples of contaminating liquids, aerosols and sprays do occur. In inorganic trace analysis the gas phase contamination route is important only in relatively few cases (cf. organic trace analysis) and particles are the major form of extraneous analyte. The source of these particles can be very varied but they are generally a mixture of particles which have been generated within the laboratory and those which are already present in the air which has been drawn into the building.

2.2.1.1 Sources

The air in the laboratory must initially come from the ambient atmosphere outside the building. Whereas the particle density in clean laboratory environments is typically in the range 1–10^5 particles per cubic foot (0.0283 m^3), in clean rural areas 10^7 particles per cubic foot are not unusual. These concentrations rise rapidly to much higher concentrations in urban environments. The particles found in urban air can be divided into three groups (Table 2.1).The particle size distribution in an

Table 2.1 Particle size ranges in urban aerosols (Whitby, Husar and Liu, 1972)

Size range (μm)	Origin and fate
0.01–0.03	Produced mainly by gas phase reactions and combustion, e.g. car exhausts containing lead, products in emissions from furnaces. Long residence time unless deposited on surfaces, washed out by rain or snow, or agglomerated to form larger particles.
ca 0.3	Produced mainly from combination of smaller particles.
5–10	Coarse particles. Liquid drops from vapour condensation or atomization. Solid particles from erosion or abrasion of surfaces. Shorter residence times in the air than smaller particles.

aged stable urban aerosol is biased strongly towards the smaller particles (Figure 2.1).

Although the finer particles are present in greater numbers, the mid-range and large particles will possess the greater volume and mass. It is therefore the larger particles that pose the greatest contamination risk. The particle concentration in ambient air outside a building will vary according to weather conditions, with a decrease in particle concentrations occurring after rain or in calm conditions, and an increase during wind events. Within buildings, the outside particle concentration will be reflected if a reasonable air exchange rate is maintained. Particle concentrations will mimic short and long term temporal variability resulting from, for example, traffic patterns and changing agricultural activities. The particle concentrations and type will of course also be dependent on location, with large differences between urban, rural or other locations, such as on a ship or ice cap.

The free exchange of air between the laboratory and the outside environment is broken in modern building design by the tendency to employ forced ventilation systems and recirculating air-conditioning systems. This will completely change the flow of particles through the structure. Air conditioners fitted with coarse filters will remove only the largest particles, and any air cleaning system in the laboratory must be able to handle efficiently the remaining particle loading in the air.

Within the laboratory, there are a range of additional sources of airborne contaminants including old painted walls, furniture, carpets and other fittings. Where trace

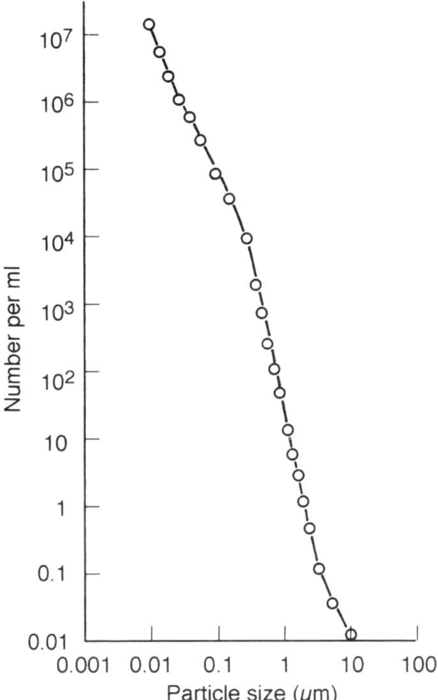

Figure 2.1 Number of aerosol particles per millimetre as a function of particle size in an aged urban aerosol (Whitby, Husar and Liu, 1972)

metal analysis is to be undertaken, metallic items are to be avoided if possible, as they can be attacked by airborne chemicals. The corrosion products, often oxides and complex carbonate compounds, are easily dislodged. Within clean working environments, the types of particle which are typically found fall into a number of categories (Table 2.2).

Table 2.2 Shape and origins of particles frequently found in clean laboratories (Lieberman, 1992)

Shape	Approximate % in dusts	Sizes (mm)	Major origins
Sphere	10	0.01–300	Liquid condensation and droplet formation
Cube	30	0.1–1000	Fragmentation of crystalline materials
Fibre	15	0.1–500	Fabrics from clothing, wipes, etc.
Flake	45	0.1–100	Skin tissue, products of metal and plastic-wear, including corroded components

2.2.1.2 Mechanisms of deposition and attachment

A range of forces acting on atmospheric particles can lead to their transport, deposition and attachment to a surface. The main forces are aerodynamic, Brownian motion, electrostatic forces and gravity (Lieberman, 1992). Particle shape, mass, size and charge can influence very significantly the relative importance of these forces. Generally Brownian motion is most important for the smaller particles, which can have long residence times in the atmosphere. However, collisions with surfaces can lead to deposition. When significant electric fields are present, charged atmospheric particles can be attracted to oppositely charged surfaces. The surface charge can be either positive or negative, and arises from a number of causes.

Charge transfer by friction is generally the most important mechanism. Others include ionization as a result of the decay of natural radioactive materials and through photo-ionization processes. The highest static charges are built up, under low humidity, on insulating surfaces that themselves have low moisture contents. The flow of air in a clean air hood, for example, can lead to a static charge building up on plastic surfaces. This is particularly a problem if the hood is closed down for periods of time, as small particles can more readily be drawn into the hood at this time in the presence of an electric field and in the absence of an air flow. The other major force influencing deposition is gravity, which increases with the mass of the particle.

Once a particle is at a surface, gravitational and inter-molecular Van der Waals forces combine to hold the particle on the surface. Electrostatic forces are believed to be of relatively minor importance. The adhesive forces operating on particles at a surface, and their relative magnitude are given in Table 2.3.

Table 2.3 Relative importance of adhesion forces for small particles at a surface in the absence of liquid films (Lieberman, 1992)

	Relative force (%)		
	1 mm	10 mm	100 mm
Van der Waals	0.4 (99)	4 (97)	40 (29)
Electrostatic	0.005 (1)	0.05 (1)	0.5 (–)
Gravity	0.0001 (–)	0.1 (2)	100 (71)
Total	0.4	4.1	140

If a particle is coated with a liquid or consists entirely of liquid, surface tension forces also act to keep the material on the surface. Unless a particle is strongly hydrophobic, increasing the humidity will increase the water adsorbed on its surface. To release a particle back into the atmosphere, the adhesive forces operating on it must be overcome by the air flow. When air flows over a surface, there is a static layer of air at the surface which is termed the boundary layer, the thickness of which depends principally on the air velocity. In clean rooms or unidirectional flow hoods, the boundary layer of static air at a surface is typically several millimetres thick at normal air velocities; thus, as the small particles do not extend out of this layer, they cannot be remobilized into the atmosphere by this flow. A detailed understanding of the processes influencing the fate of small particles at surfaces in clean room environments is still developing (see references in Lieberman, 1992).

2.2.1.3 Contaminants in atmospheric particles

The scale of the problem of particulate contamination is indicated in Table 2.4, which gives the trace metal content of deposited atmospheric dust collected from a laboratory in an ageing building.

Table 2.4 Concentrations (μg/g) of some metals leached by concentrated nitric acid from laboratory dust

Cd	Co	Cu	Fe	Mn	Ni	Pb	Zn
76	147	406	48300	393	380	2360	2400

Some elements have a much greater relative abundance, and thus contamination potential, than others. In this particular laboratory, not untypically, iron, lead and zinc levels are high; this will clearly vary greatly from laboratory to laboratory and with the contamination sources which are present. The magnitude of this source of contamination can readily be demonstrated. A typical concentration for dissolved iron in open ocean seawater is about 1 nanomolar. Thus the total iron content of one litre of such seawater is present in only about 1 microgram of this dust!

In particular cases, volatile inorganic elements and compounds pose problems due to their mobility in the gas phase. Mercury vapour, for example, can disperse readily and diffuse into unprotected reagents and samples. The presence of elemental mercury from broken thermometers or mercury electrodes used in electrochemistry does not generate a good environment for trace mercury analysis! Completely clearing spillages of mercury from inaccessible crevices can be a difficult task.

2.2.2 The Human

In all but the most sophisticated robotic analyses, human tissues and secretions are likely to come inadvertently into contact with the sample. Problems which arise from this source range from the deposition of finger lipids to the introduction of dandruff. Cosmetics can be a significant source of contaminants (Table 2.5) because of the metal oxides and other materials which are added for colour and texture. Other personal hygiene products also can give problems. Changing shampoo may lead to a drastic increase in the levels of selenium blanks, and the use of mercury based soaps (which are banned in many countries) or zinc based ointments may also lead to significant contamination. Coughing and sneezing can produce large numbers of droplets; a cough can produce 6×10^5 droplets larger than 0.5 μm which can be ejected at relatively high velocities. The skin continuously sheds dead cells and cell fragments, typically 20–40 μm in diameter and 2–4 μm in depth. Shedding of these particles will increase with sunburn and conditions which lead to drying of the skin surface.

Table 2.5 Elements present in commonly used cosmetics (Phillips *et al.*, 1983)

Cosmetic	Particles 0.5 μm or larger per application	Elements present
Lipstick	1.1×10^9	Bi
Blusher	6.0×10^8	Mg, Si, Fe, Ti
Powder	2.7×10^8	Si, Mg
Eye shadow	8.2×10^7	Bi, Si, Mg
Mascara	3.0×10^9	Fe

The number of particles emitted by individuals rapidly increases with increasing activity. Thus, an individual in normal clothing sitting or standing motionless can generate on the order of 10^5 particles (> 0.3 μm) per minute, whereas a person walking normally can generate 10^7 particles such particles per minute.

Look to yourself as a source of contamination

Not only is it difficult to decontaminate the human operative, but contamination events from this source are some of the least predictable.

2.2.3 Other Contaminants in the General Laboratory Environment

There are many other sources of contamination in an apparently clean trace analysis laboratory. Paper finds many applications in the laboratory and, although there is comparatively little contamination risk from a good quality laboratory reference or note book, the injudicious use of other paper products can cause significant contamination. Paper towelling can introduce large quantities of mobile fibres into the work place. Poor quality products are especially prone to releasing paper fibres; the most dangerous act is often breaking off sheets from a roll, as this disperses fibres into the atmosphere. High quality materials normally have a more tightly bound structure which is less prone to such releases. A range of paper and fibre based wipe products has been developed for industrial applications in clean rooms, and several of these are readily available. In particular, low fibre emitting and low dissolvable content tissues are appropriate for wiping operations in trace analysis environments. However, it must be stressed that direct contact of tissues or wipes with critical pieces of equipment or the sample itself must be avoided. In the most stringent applications a damp fibre based cloth will be preferable for cleaning and wiping up spillages, rather than tissues. There is a range of other materials commonly used in the laboratory which contain significant amounts of potential contaminants. Examples are given in Table 2.6.

> Look critically at all general purpose materials to be used in the trace analysis laboratory

Table 2.6 Examples of trace element content of materials commonly used in laboratories (taken from Robertson, 1968). Concentrations in μg/g except for Fe, which is given in mg/g

Material	Zn	Fe	Sb	Others
Paper tissues	49	1		Cr 0.5
White plastic tape	3000	67		
Adhesive tape	1.5	5		
PVC tubing	7	270	2.7	Cu 0.6
Rubber tubing	40000		0.36	Cr 420, Co 7.5
Neoprene rubber	18000			Co 3.1, Sc 3.1

2.3 CONTROL OF ATMOSPHERIC CONTAMINATION IN THE WORKING ENVIRONMENT

2.3.1 The Laboratory

The degree of contamination control required for a particular trace analysis depends principally on the mass of analyte in the sample relative to that in contamination sources, and the analytical practices used. Relatively simple precautions can dramatically reduce contamination from atmospheric sources in a conventional laboratory. Clearly the sample should be isolated from contamination sources; tops should be kept on reagent and sample containers except for the short time needed to transfer aliquots. Bottles and equipment should be kept away from the laboratory air using plastic snap-top boxes or re-sealable plastic bags. Such precautions are dealt with in the chapter on working practices. These will not totally prevent random contamination from the laboratory environment, and for most elements at low trace levels such simple steps will not be adequate. The more rigorous approach of a filtered air environment is needed.

2.3.2 Filtered Air Environments

The simplest approach is the use of filtered air enclosures. Here the air supply to an enclosure is passed through a high flow filter, typically of 0.2 μm cut-off, to

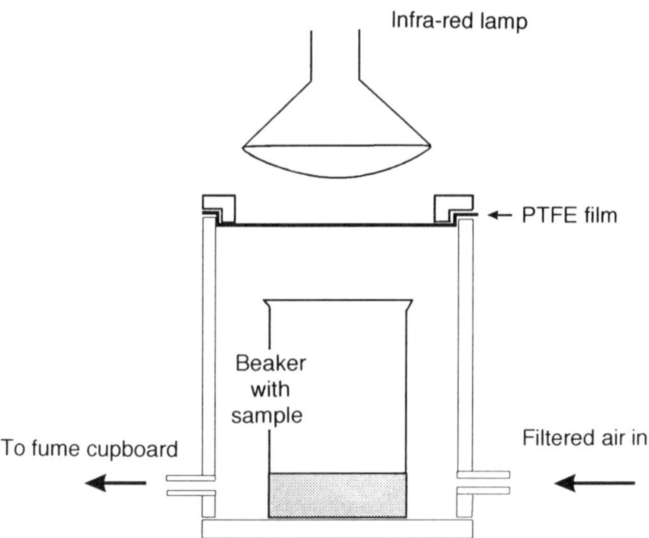

Figure 2.2 Enclosed clean evaporation system. The units are made entirely of PTFE

remove particles. The enclosure can vary in size up to typically one cubic metre, but these larger sizes are relatively inconvenient because of the remote nature of the manipulations. However, smaller units with easy access have found applications, particularly in the clean evaporation of solutions. The fume-laden exhaust can be vented to a conventional fume cupboard or extraction system. An example of this type of equipment is shown in Figure 2.2.

Enclosures are not a suitable approach if large working areas are required. The next step up is the laminar (or unidirectional) flow hood. Air is drawn through a pre-filter by a fan, and then pushed through a high efficiency particle air (HEPA) filter. The enclosure and HEPA filter are arranged to provide a rapid laminar flow through the enclosure, thus ensuring that the whole hood is purged by the filtered air. There are several variants on this basic design. In some, a fraction of the HEPA filtered air is drawn through holes in the work surface and is recirculated through the HEPA filter. A typical system is shown in Figure 2.3.

Although the term laminar flow is normally used to describe the air flow within these devices, it has been pointed out (Lieberman, 1992) that it is more appropriate to refer to the flow as unidirectional, as any obstruction caused through work in the enclosure will disrupt the ideal flow regime. A compact version of a non-unidirectional filtered air environment for handling small samples is a plastic sphere with cut-outs for hand and sample access; air is pushed by a fan through a small HEPA filter and into the sphere.

The relative efficiencies of the filter media used to clean air introduced into hoods and laboratories is shown in Table 2.7.

Table 2.7 Filter media and their removal efficiency for particles 0.3 μm diameter or greater

Filter	Removal efficiency
Pre-filter	80–95%
High Efficiency Particulate air (HEPA) filter	> 99.97%
Ultra low penetration air (ULPA) filter	> 99.997%

The HEPA filter consists of a random mat of fibres about 0.1 μm in diameter. It is important to realise that the sieving action of this mat plays a minor role in the removal of airborne particles. The average aperture size in the matrix is often significantly greater than 0.5 μm, which gives a high air flow and minimum pressure drops across the filter. The most important removal processes are those which lead to the collection and retention of particles at surfaces. These retaining forces have been discussed above; the processes which lead to the deposition are:

(1) interception, where the particle in a stream of air is brought very close to the surface of a fibre so that it can be retained;

(2) impaction, where the momentum of a particle in an air stream is adequate for it to directly impact the surface of a fibre (important for larger particles);

(3) for very small size classes, diffusive processes which can bring the particle to the surface.

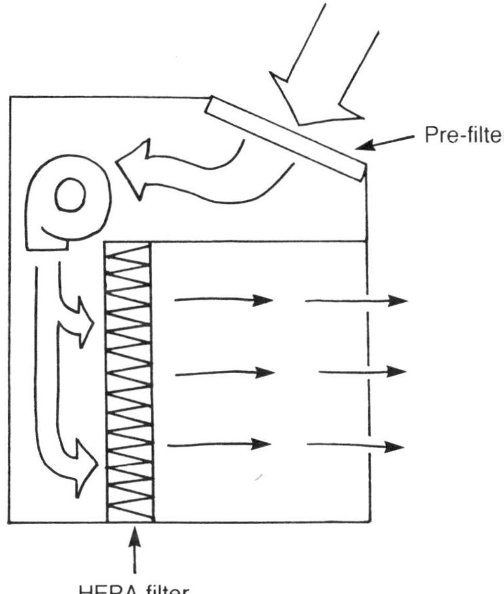

Figure 2.3 A schematic cross-section through a unidirectional flow hood

The filter medium is normally in a pleated format with dividers, to provide a large surface area in compact dimensions. Often these filters have protective grids across their face made of metals such as anodized or painted aluminium. In conventional clean room applications these may not produce particles. The problem arises if vapours of aggressive chemicals are drawn through the filter, as these may attack the grid. This will also occur if reagents are left in closed, non-operating hoods.

The degree of filtration of the air falls into set standards based on the number of particles of particular size per unit volume of air. Probably the most commonly used standard has been the FS209D standard from the USA. This is most usually thought of in terms of the number of particles greater than 0.5 μm diameter per cubic foot of air (0.0283 m^3). The full standard, however, provides maximum permissible numbers of particles in smaller size classes (Table 2.8) as well as appropriate air sample volumes for valid particle counting. Other commonly used standards from other countries are compared with the FS209D standard in Table 2.9.

Table 2.8. Federal Standard 209D air cleanliness Class levels. The numbers shown are the maximum number of particles greater than the specified size (in μm) permissible per cubic foot (0.028 m^3) of air. Note that the numbers of particles shown do not make any assumptions about the actual size distributions of particles to be found in particular situations

	Particle size (μm)				
Class	0.1	0.2	0.3	0.5	5.0
1	35	7.5	3	1	NA
10	350	75	30	10	NA
100	NA	750	300	100	NA
1000	NA	NA	NA	1000	7
10000	NA	NA	NA	10000	70
100000	NA	NA	NA	100000	7000

NA = not applicable.

It is very important to be aware that such standards and applications of laminar flow equipment and clean rooms are generally *not* specifically designed for use in trace analysis. Although some features may be highly appropriate, others can potentially be sources of contamination, and meeting a specific air standard does not guarantee good trace element analyses.

The materials used in the construction of laminar flow hoods need to be carefully assessed for compatibility with trace analysis procedures. Metallic components which will corrode if aggressive chemicals are used, such as aluminium dividers for the pleated HEPA filter, metallic screws, and metallic screens to protect the HEPA filter, should either be replaced or coated with lacquer or epoxy paint. The motor and fan assembly are also of course metallic, and appropriate parts can be coated. However, as these items are upstream of the HEPA filter, they will be a major source of contamination only if significant numbers of particles less than 0.5 μm in diameter are generated. Whilst there should be an awareness of this source, no problems in this area have been noted in the literature.

Table 2.9 A comparison of national standards for major classes of clean room; bracketed date is year of issue of standard (Whyte, 1991)

USA 209D (1988)	Australia (1989)	Britain (1989)	France (1981)	Germany (1990)
				0
1	0.035	C		1
10	0.35	D		2
100	3.5	E or F	4000	3
1000	35	G or H		4
10000	350	J	400000	5
100000	3500	K	4000000	6
		L		7

As with all other aspects of trace analysis, care must be taken with the use of laminar flow hoods. If a particle-generating object is placed in the unidirectional flow, the area down stream will be contaminated. Thus, the location and order of processing of items in the hood should be considered in the most rigorous applications, and clearly dirty and particle-generating items should be excluded from the hood. Under some circumstances it is possible for particles to enter the hood, such as when the hood is located close to a door, and the air movement caused by a rapid opening can lead to turbulence in the air stream with a net transport of material into the work area. Sensible location of the hood is an obvious cure.

Where toxic or hazardous materials are to be used, the conventional unidirectional flow hood (UFH) has a major drawback in that all fumes generated in the working area are swept back into the face of the operator. Additionally, the problem of corrosion and deterioration of surfaces and components within laminar flow hoods must be kept in mind. There may be reagents stored within a UFH, and acidic vapours can lead to the eventual corrosion of components, even when coated with paint or plastic. In the Southampton laboratory high aluminium blanks were traced to the deteriorating paint coating on the protective aluminium grid of a HEPA filter. A safe environment for the transfer of acids and other volatile noxious reagents, clean evaporations, and related operations, which still has a controlled particle environment, is the clean wet station. Here HEPA filtered air flows vertically down onto the work area, and is drawn down through holes or ducts in this surface, carrying any noxious fumes with it, into a conventional fume cupboard vent system. Scrubbing can be incorporated if required to clean the exhaust gases. When the analysis dictates the use of relatively sophisticated equipment such as clean wet stations, the analysis is moving to the situation where a dedicated clean laboratory is required.

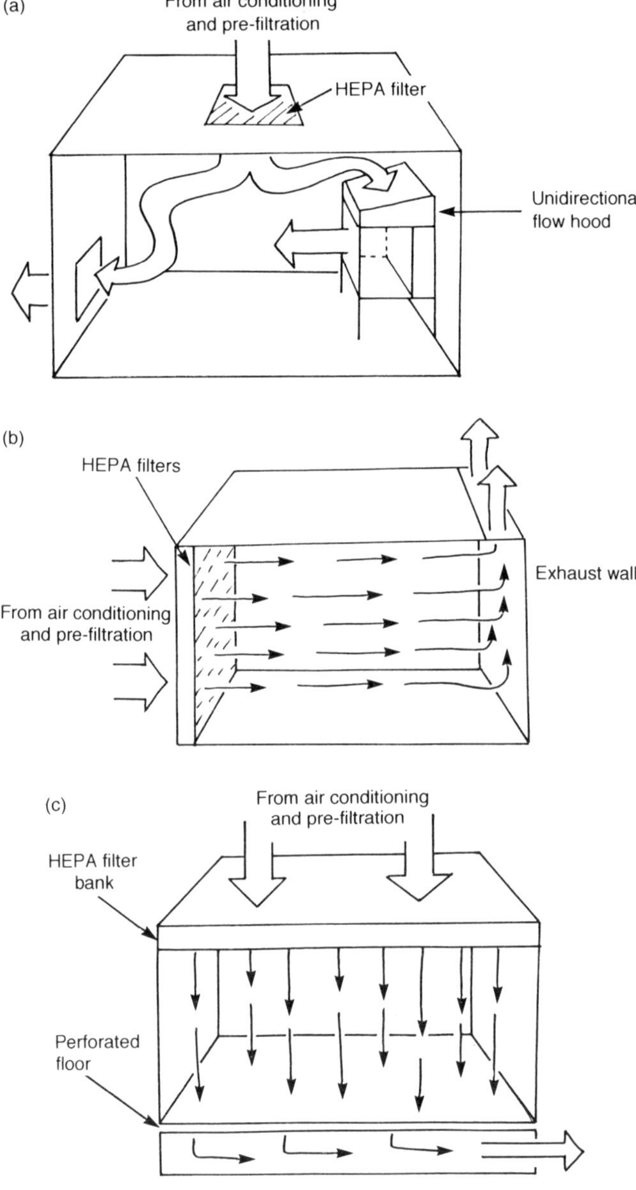

Figure 2.4 Schematic diagrams of different configurations of clean room, showing air flows.
(a) Non-directional flow of filtered air to clean room, with individual unidirectional flow hoods within room.
(b) Horizontal unidirectional flow clean room. Air becomes more contaminated with particles as one moves away from the bank of HEPA filters.
(c) Verticle unidirectional flow clean room. There is uniform high quality air across horizontal planes in the room, with the highest quality closest to the HEPA filters. This type is used for most stringent applications when humans are present

2.3.3 The Clean Laboratory

The next phase in improving a working environment in which particulate contamination is controlled is to use a full clean laboratory. A good general account of cleanroom design is given in the book edited by Whyte (1991). In modern clean laboratories, the materials used, laboratory construction and the filtration of all air introduced provide a strict control of airborne particulate contamination. Most clean rooms in industrial applications will have at least an anteroom to isolate the fully filtered room from the general laboratory environment and to provide an area for changing into clean room clothing. It may also have an air shower to remove loosely held particles from garments. The changing room is often separated from the clean room itself by a raised barrier as a physical reminder of the transition into a clean area. A hatch system flushed with filtered air can be installed, to allow the transfer of items between the clean room and the outside world without the analyst having to change clothes and leave. More sophisticated installations may have a series of rooms in which one progresses from, for example, Class 10 000 to Class 100 and then on to Class 10 conditions. A basic concept behind the operation of such rooms is that they are under positive pressure to prevent unfiltered air entering.

In the most rigorous industrial applications, vertical down-flow rooms are used, to generate better than Class 10 conditions. Here the entire ceiling is a bank of HEPA filters, and a unidirectional flow of filtered air passes down to a perforated floor; the air with any particulate contamination is drawn away through a plenum below the floor [see Figure 2.4(c)]. The next step down is the horizontal flow clean room, where a wall consists of a bank of HEPA filters. In the vicinity of the wall Class 100 conditions will be present, but as the air flow proceeds to the receiving wall at the opposite end of the laboratory the air stream will pick up particles and becomes increasingly contaminated [Figure 2.4(b)]. In less critical work, non-unidirectional HEPA filtered air can be provided to a laboratory through ducts, and critical work within this environment can be done within unidirectional flow hoods [Figure 2.4(a)]. There is a very large difference between the extremes in terms of installation and maintenance costs, with the vertical flow systems being the most expensive.

There are several important constructional features of general clean rooms. Surfaces must be non-particle-generating, easily cleaned, and have no sharp corners or crevices in which particles and contaminants can lodge. The meeting edges of ceiling, floor and walls are thus usually coved. Objects which will disrupt the unidirectional flow are minimized, and critical equipment is sited in the cleanest location within the room.

2.3.3.1 Application of clean rooms to trace analysis

Clean rooms have been most often used in industrial applications such as the production of pharmaceutical, precision engineering, and micro-electronic products. Although there is obvious overlap in criteria, industrial clean rooms do not usually reflect the detailed requirements of a laboratory set up for trace element analysis. As with other aspects of ultra-trace inorganic analysis, attention must be focused on the most critical aspects of the construction and use of the room as regards contamination. There are thus a series of additional points about the construction of clean rooms which are specific to trace element analysis.

Metallic components should generally be avoided, because of the potential for corrosion and particle generation. Screw heads, sink fittings, plumbing components of brass, copper and lead and exposed constructional members are all examples of items that should be excluded from the design, or if essential should be isolated from the general laboratory environment, or coated with epoxy paint or similar corrosion resistant material. The ducting used in the air handling systems for clean rooms is frequently zinc coated. Although this lies upstream of the final HEPA filters, it represents a potentially huge contaminant reservoir of this element; coated (e.g. rubberized solution) or solid PVC sections are to be preferred. The material of choice for fume ducting on clean wet stations is also PVC. If ducting is found to be a source of contamination after construction, a large cost is involved in modifications.

At the most basic level, an effective clean environment can be a polyethylene tent suspended within a rigid framework, to which filtered air is supplied from a Class 100 laminar flow module. A separate Class 100 laminar flow hood recirculates the air in the plastic tent until it is vented from the enclosure. The framework may have a purpose built door or overlapping sheets of plastic. Whereas this type of approach has been used, for example, for research expeditions on ships, and represents a very cost effective way of providing a clean enclosure, it has long term problems. There is no temperature or humidity control, cleaning of the flexible surfaces is difficult, and generally maintaining the clean integrity of such systems over long periods is difficult. A more substantial facility, with its large installation and running costs, is really needed.

Table 2.10 Clean laboratories used during trace inorganic analytical operations

Application	Reference
Fixed laboratories	
Lead in the environment	Patterson and Settle (1976)
Trace metals in seawater	Gardner (1979)
National Bureau of Standards trace element laboratory	Moody (1982)
Trace elements in ice cores	Boutron and Batifol (1985)
Transportable clean laboratories	
Collection of clean sea water samples	Wong *et al.* (1977)
	Danielsson and Westerlund (1983)
	Morley *et al.* (1988)

A good account of a clean laboratory built specifically for trace element analysis at the National Bureau of Standards (now the National Institute of Science and Technology), and containing much useful detail, is given by Moody (1982). Here the features of a downflow clean room have been cost effectively married to a laboratory environment specifically developed and tested for trace inorganic analysis applications. Various clean laboratories for use in trace analysis are described in the literature, and a selection is listed in Table 2.10.

The importance of maintaining the integrity of samples during collection and handling away from the laboratory has led to the development of portable clean laboratories which are designed to be used in the field or at sea; examples are also given in Table 2.10.

2.3.3.2 Maintenance of clean facilities

As with any complex facility, clean laboratories, and to a lesser extent clean unidirectional flow hoods, require regular and thorough maintenance. Recurrent costs of materials and technical time should be built in to any forward financial planning involving the use of clean facilities. Periodic replacement of HEPA and pre-filters is essential, and the need for replacement can usually be determined by the pressure drop across the filter as it becomes progressively blocked by particles. If this operation involves the replacement being carried out within the clean area, then the facility will need careful cleaning of e.g. surfaces, light covers and key equipment before any analysis is done.

A pro-active attitude of the analysts who work in the clean environment is essential. There should be a continual awareness of problems which may be developing, e.g corrosion of a component or a build-up of unused materials and equipment in the laboratory, which may lead to subsequent contamination problems. It is much better to deal with such problems at an early stage than to find high and variable blanks later, which invalidate a series of analyses which have involved time and resources to undertake. To initiate this considerable awareness there needs to be training of the analysts involved.

2.3.4 Cleaning Procedures for Surfaces

As indicated earlier, once particles are deposited at a surface, they are not readily moved by the ambient air flows in a clean hood or laboratory. However, if critical pieces of equipment or samples are brought into contact with the surface, transfer of some contaminant can occur. Appropriate cleaning is therefore important. In a comparison of different cleaning methods (Lieberman, 1992), which included sophisticated high pressure washing, ultrasonic agitation, and use of hydrophobic solvents such as Freon TF, the most effective method for removing particles was the simple procedure of wiping with a lens tissue using ethanol and acetone as the solvents. In the Southampton clean laboratory wiping with ethanol using a non-fibre-generating tissue has also been found to be a simple and effective way to clean plastic surfaces. Ethanol is preferred to acetone because of the lower potential health hazard and the lower absorption into plastics. In many cases high purity water can also be an effective wash agent.

The elimination of static electricity is an important aspect of cleaning surfaces. The use of proprietary static eliminators can help, as does the routine wiping of surfaces with ethanol, which will disperse such charges as well as clean the surfaces.

2.3.5 The Human

Many of the sources of contamination identified earlier in this chapter can be limited by appropriate measures. However, the most important principle is that there must be an awareness of the problems by the analyst. This means there must be training on contamination control measures for individuals as an important component in any overall contamination control programme. A continuous critical appraisal of working practices is needed for successful control, both by the researcher and if possible others with an awareness, but an independent view, of the problems involved.

Widely varying levels of precautions are taken to control this problem. When the analyst is a source of potential contamination, the obvious first order answer is to isolate him or her from the samples with appropriate clothing. The most fundamental precaution is to prevent contamination from hands by wearing gloves, as virtually all analytical procedures involve some manipulations by the analyst. Both pigmented and rubber based gloves should be avoided if possible because of inorganic filler materials, catalysts and other materials in the matrix of the plastic. The use of talc inside some brands, which allows easy entry of the hand, is another major potential source of contamination. Thin clear disposable polyethylene gloves provide a generally acceptable alternative. These can normally be used without prior cleaning, and comfort can be improved by wearing glove liners made from non-particle-generating fabric. It is important to remember that the opening of an outside door, or the picking up of a tool or other source of analyte, makes the glove a contamination source.

Gloves are only as clean as the last thing touched

Gloves must therefore be routinely changed to take account of their use in potentially contaminating operations. Gloves are cheap in comparison to the cost of collecting and processing samples. Polyethylene gloves may be clean but they are mechanically fragile. Their advantages must be weighed against safety aspects, and heavy duty acid resistant gloves will be necessary for removing items from acid cleaning tanks if no other cleaner means, such as plastic tongs, are available. Under these conditions contaminants should be in solution, and so, after an initial rinse with water to remove excess acid, a final rinsing with high purity water can be done wearing gloves with less contamination potential. Guidelines for appropriate constructional materials for gloves used in different procedures are given in Table 2.11.

Table 2.11 Appropriate materials for gloves for different handling procedures in trace inorganic analysis. *Note*: Always check the behaviour of gloves with reagents before use on hands. Always rinse gloves with water after use

Glove material	Typical applications
Neoprene/heavy duty PVC	Insertion and removal of equipment from acid tanks; dispensing bulk acids and solvents. For protection of analyst
Natural rubber; latex	For use with less concentrated acids. *Note*: permeable to a range of aromatic and aliphatic solvents. For protection of analyst
Polyethylene film	Handling of critical pieces of equipment; for protection of apparatus and samples from analyst

The types of garment used in clean rooms in the micro-electronics industry provide a good model and source for many trace analytical applications. Laboratory coats and other clothing should be:

(1) made from materials that do not shed fibres,

(2) with all cut edges enclosed,

(3) made without components that can corrode or contaminate (e.g. metal buttons and zips),

(4) able to retain particles (i.e. not open weave, as these may allow transmission of particles),

(5) comfortable to wear, and if possible resistant to acids and reagents.

Materials used include polyester fabrics, spun bonded nylons, and polypropylene composites. For more critical work, head covers and beard/moustache covers ('snoods') should be worn. A variety of cosmetic, clinical and cleaning products used on hair and skin can be significant sources of contamination, as discussed above. Footwear can transport significant amounts of debris into a clean environment. This is normally controlled by use of overshoes and tacky mats positioned in front of clean laboratory entrances which will remove particles from the bottom of clean shoes.

In the most rigorous applications, such as the measurement of trace metals in ice cores or lead in deep ocean water, full clean laboratory clothing is worn. This will typically consist of either a one piece suit or trousers and coat together with overshoes, and a covering for the head and, if appropriate, the beard or moustache. These garments are put on in an anteroom, so that conventional clothing does not enter the clean area. Table 2.12 gives guidelines for the clothing appropriate to particular clean room environments. Whilst these protocols are based on microelectronics industry procedures, and therefore may require some modifications for successful use in the clean analytical laboratory, they provide useful initial guidelines for work in trace analysis.

Table 2.12 Guidelines for clothing protocols in clean environments

Class of room/facility	Appropriate clothing
Unidirectional flow hood in conventional laboratory	Laboratory coat, polyethylene gloves; i.e. all parts of body extending into the hood must be covered by non-contaminating clothing
Class 100 000 room	Minimal specialized clothing; smocks, street shoe covers, head covers
Class 10 000 or 1000 room	Complete clean room uniform with gloves, breathing zone cover; no cosmetics or rapid body movements
Class 100 or 10	Complete change of clothing; special gloves and foot covers, minimal contact with critical equipment
Below Class 10	Robotic systems; no humans if possible

The extent to which precautions are needed will ultimately depend on the element(s) of interest and the level and reproducibility of blanks. In all cases, if the precautions are to operate effectively, there must be a continual awareness of the potential problems stemming from the presence of the analyst, and regular and effective protocols must be maintained. Future trends in trace analysis will involve the increasing isolation of the analyst from the analytical operations. This might be by the use of in situ analysers, such as the in vivo monitoring of blood constituents. Alternatively, closed systems with more readily controllable environments may be employed. This latter approach is becoming widely adopted in the micro-electronics industry.

2.4 HOW CLEAN DOES YOUR ENVIRONMENT NEED TO BE?

Clearly the use of a full clean room for processing samples represents a very rigorous approach, and some groups have argued that for very difficult analyses, such as lead in seawater, this is the approach that must be used for accurate analyses. However, other groups have attained excellent results with much more modest facilities, but with very careful attention to the critical handling and analytical steps, and argue that a full clean room is not necessary. Generally speaking it is much easier to control random contamination from the laboratory environment if that environment is clean.

Another factor is cost. The full vertical flow clean laboratories are extremely expensive to build and run. A much more economical approach for trace analysis is to have a non-unidirectional flow of filtered air into a room and then to have high quality clean hoods for critical work such as transferring samples and key chemical processes.

Keep critical steps to restricted high quality areas

This approach has been used very successfully by several groups involved in the low level analysis of environmental samples.

Ultimately, the choice of working environment for the elements of interest will depend on the handling and analytical sequence used, the magnitude and variability of the blank, and how accurate and precise the data are as a result of these factors. At very low concentrations the question of accuracy is rarely straightforward. Unfortunately there is no simple list of steps to follow to guarantee insignificant contamination from the working environment for particular analyses. Individual work areas and analytical requirements can vary greatly, and it must be the quality of the analyses that will indicate the level of precautions needed. The logical sequence is to start with the simplest system and upgrade as required, using blank determinations and certified reference materials for rigorously assessing the data in the quest for reproducible and accurate analyses.

REFERENCES

Boutron, C.F. and Batifol, F.M. (1985) Assessing laboratory procedures for the decontamination of polar snow or ice samples for the analysis of toxic metals and metalloids. *Ann. Glaciol.* **7**, 7.

Danielsson, L.-G. and Westerlund, S. (1983) Clean room container laboratory for YMER 80. *Ocean Sci. Eng.* **8**, 53–62.

Gardner, D. (1979) Design and operation of a laminar flow cleanroom for trace metal analysis in marine chemistry. Report 107, Commonwealth Scientific and Industrial Research Organisation, Division of Fisheries and Oceanography, 19 pp.

Lieberman, A. (1992) *Contamination control and cleanrooms: problems, engineering solutions, and applications*. Van Nostrand Reinhold, 411 pp.

Moody, J.R. (1982) The NBS clean laboratories for trace element analysis. *Anal. Chem.* **54**, 1358–1376.

Morley, N.H., Fay, C.W. and Statham, P.J. (1988) Design and use of a clean shipboard handling system for seawater samples. *Advances in Underwater Technology, Ocean Science and Offshore Engineering*, **16**, 283–289.

Patterson, C.C. and Settle, D.M. (1976) The reduction of orders of magnitude errors in lead analysis of biological materials and natural waters by evaluating and controlling the extent and sources of industrial lead contamination introduced during sample collection, handling and analysis. In: Trace Analysis: Sampling, Sample Handling and Analysis, NBS Special Publication No. 422. National Bureau of Standards, Washington DC.

Phillips, Q.T. Auser, W.D. Baldwin, J.M. and Washington, G.J. (1983) Cosmetics in clean rooms *J. Env. Sci.* **26**(5), 27–31.

Robertson, D.E. (1968) Role of contamination in trace element analysis of sea water. *Anal. Chem.* **40**, 1067–1072.

Whitby, K.T., Husar, R.B. and Liu, B.Y.H. (1972) The aerosol size distribution of Los Angeles smog. In *Aerosols and Atmospheric Chemistry,* G.M. Hidy (Ed.), Academic Press, New York, pp. 237–264.

Whyte, W. (Ed.) (1991) *Cleanroom Design*. Wiley, Chichester.

Wong, C.S., Cretney, W.J., Piuze, J., Christensen, P. and Berrang, P.G. (1977) Clean laboratory methods to achieve contaminant-free processing and determination of ultra-trace samples in marine environmental studies. In Methods of Standards and Environmental Measurement. NBS Special Publication No. 464, National Bureau of Standards, Washington DC.

3 Laboratory Materials

Everyday items of laboratory equipment such as pipettes, beakers, storage bottles etc. play an important role in all analyses. Because these common items perform well in most conventional analyses, it is often assumed that conventional laboratory apparatus will be suitable for trace analysis. Everyday items, however, have become such because many analytical applications are not critical. A more positive attitude must be taken if such equipment is not to introduce complications into a trace analysis.

3.1 CHEMICAL CONSIDERATIONS

Classical chemical analysis procedures suffer from only minimal contamination and loss problems. It has therefore been possible to develop apparatus made from readily available materials such as glass, and to place an emphasis on quantitative transfer of the analyte to minimize sample losses. In a trace analysis, however, it must not be assumed that such apparatus is appropriate; it may either introduce trace contaminants into the sample or present an active surface onto which the analyte may be adsorbed, and hence lost from solution. The choice of material for the manufacture of a particular piece of apparatus will rarely have taken into account its potential to cause contamination and losses in a trace analysis. Great care must therefore be taken in assessing the potential problems which might be encountered.

In trace analysis the analyte is present in such low concentrations that the surface of a vessel may present sufficient active sites for the sorption of the analyte by processes such as ion-exchange. The microscopic roughness of the surface may in addition provide nucleation centres for the precipitation of materials from solution.

The best material for the construction of apparatus will vary from element to element and with the analytical sequence to be used. If the chemical basis of the analytical procedure is understood and a positive approach is taken to the choice and preparation of materials employed in the analysis, many of the all too frequently encountered contamination and loss problems can be avoided. The first step in preventing such problems is to understand the chemical nature of the materials employed in the construction of laboratory apparatus and how they might interfere in the analytical procedure. Ideally one inert high purity construction material would be suitable for all applications, but the physical and chemical requirements of the diverse range of inorganic trace analyses makes this impossible. It is therefore necessary for the analyst to understand how, and from what, his apparatus is constructed.

3.2 PHYSICAL CONSIDERATIONS

The physical properties of a construction material must be considered with its chemical properties. A material which contaminates the sample will be unusable, but problems can also occur if it has poor physical properties. Gas permeability through a material can in itself lead to contamination or sample loss if the analyte is able to permeate into or out of the container. Problems may also result if the analyte is susceptible to oxidation and the container does not fully exclude air. Less subtle problems may occur if the container fractures at low temperature and is to be employed for the storage of frozen samples.

3.3 GLASSES

The most commonly encountered material for the construction of laboratory apparatus is glass. Although appearing to be inert in most applications, this material has a highly active surface typified by a surface structure made up of acidic Si—OH groups (Figure 3.1).

The overall purity of the material, and hence its potential for contamination, is at least in part derived from the main constituents which are used in its manufacture (Table 3.1).

These are not the only elements present. Elements may be deliberately added during manufacture or result from impurities in the raw materials. With the exception

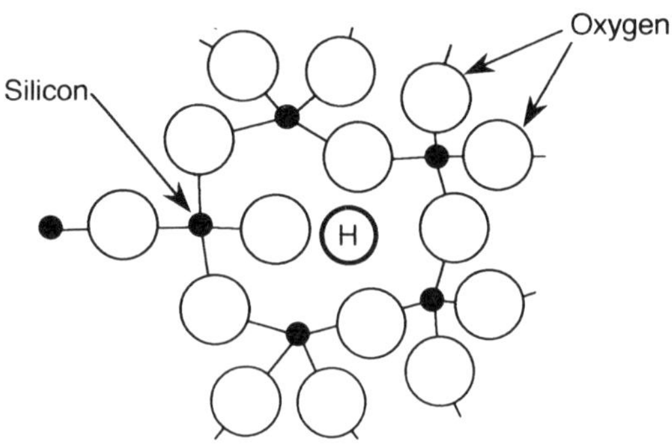

Figure 3.1 The main structural features of the surface of glass

of the silicas, species such as fluorine, chlorine, sulphate, arsenic and antimony may be present at concentrations between 0.05 and 0.5%.

Table 3.1 The major chemical constituents of glass (Adams, 1972)

	Typical concentration (wt%)									
Type	SiO_2	Al_2O_3	ZrO_2	Na_2O	K_2O	Li_2O	B_2O_3	CaO	MgO	BaO
Soda A	73	1		17	0.5			5	4	
Soda B	74	2		13	0.5		3	11	0.5	
Borosilicate A	81	2		4	0.5		13			
Borosilicate B	73	6		7	0.5		10	1		2
Alkali-resistant	71	1	15	11	0.5	1				
High silica	96	0.5								
Vitreous silica	100									

Borosilicate glasses are now used for the construction of a high proportion of the glass vessels encountered in the laboratory. These are much more chemically and physically robust than soda glass, but still release trace elements in a number of ways. Particularly severe problems occur at both extremes of acidity. High pH results in the dissolution of the glass surface itself, whereas low pH favours metal dissolution and desorption from the glass surface. As might be expected, the elements from which the glass is made pose particular contamination problems, with sodium and aluminium from soda and borosilicate glasses being particularly troublesome.

Contamination from the glass is not the only potential hazard which can be encountered in its use. The clean glass surface is chemically highly reactive, primarily because of the presence of Si—OH groups (Figure 3.1). These groups are acidic in nature and are an effective ion-exchange material. Cationic analytes are therefore efficiently scavenged from solution onto the glass surface.

$$\text{GlassSiOH} + \text{M}^+ \rightleftharpoons \text{GlassSiOM} + \text{H}^+$$

The number of such exchange sites varies from glass to glass and is particularly affected by the history of the particular vessel. Heat treatment results in the cross-linking of adjacent SiOH groups, reducing the ion-exchange activity of the surface. Treatment of the glass surface with acid or alkali, on the other hand, results in etching of the surface and the production of a new active surface.

A much better, although more expensive, choice is vitreous silica. There are several types of silica available and these differ in their applications in trace analysis. Vitreous silica can be manufactured from sand or natural quartz or by the vapour phase hydrolysis of silicon compounds such as the tetrachloride or tetrafluoride. The latter source of silica, coming from a gaseous silicon source, is significantly purer than that obtained from natural material (Table 3.2).

Table 3.2 Trace impurities (ppm) in vitreous silica (Hetherington and Bell, 1972); nd = not detected

	Silica		
Element	Transparent from quartz	Transparent from SiX_4	Translucent from silica sand
Al	74	<0.25	500
B	4	0.1	9
Ca	16	<0.1	200
Cr	0.1	0.03	nd
Cu	1	<1	nd
Fe	7	<0.2	77
K	6	0.1	37
Li	7	nd	3
Mg	4	nd	150
Na	9	<0.1	60
P	0.01	<0.001	nd
Ti	3	nd	120

Vitreous silica is an inherently clean material which can be further rigorously cleaned, and is stable at high temperatures with an essentially zero (0.54×10^{-6}) coefficient of expansion. However, it is more difficult to work than glass and is both expensive and fragile. It has excellent transmission characteristics over the range 160 nm to 4 μm. The low wavelength cut-off is largely governed by the presence of metal impurities in the silica and the highest purity grades therefore have the best transmission characteristics in this region.

3.4 PLASTICS

A wide range of plastics has now been used in the construction of laboratory apparatus, and careful selection of materials for trace analysis applications is necessary if apparatus is not to interfere in the procedure. There are a number of features which must be considered when selecting suitable materials. Some are of a chemical nature, such as the adsorption properties of the polymers and contamination arising from residual catalysts and additives used in manufacture. Other considerations may be of a more physical nature such as whether the containers can be dried at elevated temperatures. In order to assist in the selection of suitable container materials, it is helpful to know the structure and properties of the materials involved. The following summarizes some of the most important properties of commonly encountered polymers. Information is also given on some of the chemicals which are used in their manufacture. It should be noted, however, that manufacturing processes are often closely guarded secrets and the exact manufacturing process used in the polymer production and its subsequent moulding can rarely be ascertained. Manufacturers use different catalysts and fillers in their products and container manufacturers often change the supplier of their polymers. In addition, the manufacture of containers itself can lead to trace metal contamination from processing chemicals such as molybdenum sulphide, tungsten- and copper-loaded lubricants and silicones. Tin and nickel stabilizers are used in a variety of polymers. The manufacture of bottle caps in particular requires the incorporation of solid fillers such as carbonates to provide strength. The cap is frequently not itself in contact with the container contents, contact being restricted by a cap liner. Replacing the supplied liners with PTFE equivalents can sometimes be an effective solution. In general it is better to avoid all coloured plastics (including those which are white rather than translucent) as many of the colouring materials are major sources of contamination.

3.4.1 Types of Plastics

3.4.1.1 Low density poly(ethylene) (LDPE)

$$—[—CH_2—CH_2—]_n—$$

This polymer is formed by the high pressure polymerization of ethylene, and its relatively low density arises from approximately 2% chain branching giving rise to an open structure. LDPE is generally unreactive at room temperature but it is slowly attacked by strong oxidizing agents and softened or swollen by some solvents. It can be used continuously at 80 °C, or for short periods at 95 °C. Few processing metals are employed in the polymer manufacture and LDPE should not be a major source of contaminating metals.

3.4.1.2 High density poly(ethylene) (HDPE)

$$—[—CH_2—CH_2—]_n—$$

This is an unbranched linear polymer produced catalytically from ethylene. Its structure is more closely packed than that of LDPE and therefore has a higher density and greater chemical resistance. It is harder and more opaque than LDPE and can be used up to 110 °C continuously or at 120 °C for short periods.

The manufacture of HDPE is a catalytic process potentially involving some of the following elements: aluminium, chromium, silicon, titanium, vanadium, zinc and zirconium. The contamination risk may therefore be very high.

3.4.1.3 Poly(propylene) (PP)

$$—[—CH_2—\underset{\displaystyle CH_3}{\underset{|}{CH}}—]_n—$$

Poly(propylene), which is prepared catalytically from propylene, possesses a methyl side chain which makes it more susceptible to attack by strong oxidizing agents than HDPE. It can, however, be used at higher temperatures; 140 °C for short periods and 130 °C continuously. It is therefore well suited to applications requiring sterilization. PP manufacture is similar to that of HDPE, using similar catalysts and therefore posing a potentially severe contamination risk for some analyses.

3.4.1.4 Poly(methylpentene) (PMP)

$$\text{—[—CH}_2\text{—}\underset{\underset{\displaystyle \text{H}_3\text{C—}\underset{\displaystyle\text{H}}{\underset{|}{\text{C}}}\text{—CH}_3}{|}}{\underset{\displaystyle\text{CH}_2}{\underset{|}{\text{CH}}}}\text{—]}_n\text{—}$$

PMP, which is similar to poly(propylene) but with an isobutyl group replacing the methyl, is a highly transparent rigid plastic which is particularly well suited to the construction of volumetric apparatus. It withstands temperatures of 200 °C for short periods and 180 °C continuously but, as with PP, the polymer is subject to attack by strong oxidizing agents. Some solvents (such as trichloroethylene) may cause softening and swelling. The same range of catalysts is employed as with HDPE and the contamination risk is therefore high.

3.4.1.5 Poly(styrene) (PS)

$$\text{—[—}\underset{\displaystyle C_6H_5}{\underset{|}{\text{CH}}}\text{—CH}_2\text{—]}_n\text{—}$$

Polystyrene is a hard, rigid polymer which is transparent, with good dimensional stability, and which is used in a number of disposable products. It is stable to attack by many aqueous solutions but is soluble in aromatic and halogenated solvents. It is rated for a maximum temperature of 60 °C in continuous use.

When PS is copolymerized with acrylonitrile to give the styrene—acrylonitrile copolymer SAN, rigidity, impact strength and chemical resistance to non-polar solvents and aromatics are improved. The graft polymer ABS (acrylonitrile—butadiene—styrene) has a higher chemical resistance than PS and retains its shape up to 100 °C.

3.4.1.6 Polycarbonate (PC)

$$\text{—[—O—C}_6\text{H}_4\text{—}\overset{\displaystyle CH_3}{\overset{|}{\underset{\displaystyle CH_3}{\underset{|}{\text{C}}}}}\text{—C}_6\text{H}_4\text{—O—}\overset{\displaystyle O}{\overset{\|}{\text{C}}}\text{—]}_n\text{—}$$

Polycarbonate, a linear polycarbonic acid ester with good dimensional stability and impact strength over a wide temperature range, is often used in the construction of laboratory apparatus such as safety shields and vacuum desiccators. It can be autoclaved, but this can result in some loss of strength and should not therefore be carried out on such apparatus as vacuum desiccators. The polymer is subject to attack by strong acids and bases and is soluble in a number of organic solvents.

3.4.1.7 Poly(methyl methacrylate) (Acrylic)

$$\text{—[—CH}_2\text{—}\overset{\overset{\displaystyle COOCH_3}{|}}{\underset{\underset{\displaystyle CH_3}{|}}{C}}\text{—]}_n\text{—}$$

This highly transparent rigid plastic is resistant to inorganic acids and alkalis but is attacked by many organic solvents. Its main uses are in sheet form as tanks and trays and in the manufacture of burettes. It can be used at temperatures up to 70 °C continuously and at 90 °C for short periods.

3.4.1.8 Poly(vinyl chloride) (PVC)

$$\text{—[—CH}_2\text{—CHCl—]}_n\text{—}$$

PVC is structurally similar to poly(ethylene) but with a chlorine replacing a hydrogen on alternate carbons. PVC is itself hard and rigid and has to be made soft and pliable for applications such as tubing manufacture by the addition of plasticizers such as phthalate esters. It is attacked by many organic solvents and should not be used above 70°C except for short periods. It has a low gas permeability. Generally described as a dirty impure polymer, it may contain cadmium- or barium-based stabilizers.

3.4.1.9 Poly(tetrafluoroethylene) (PTFE)

$$\text{—[—CF}_2\text{—CF}_2\text{—]}_n\text{—}$$

The chemical resistance of PTFE is exceptional. It is incombustible and insoluble in all known solvents and is attacked only by molten alkali metals and by fluorine at high temperatures. It is neither distorted nor swollen by organic solvents but can be permeable to some gases, solvents and acids.

PTFE can be used at up to 300 °C for short periods and continuously at 260 °C. Its low coefficient of friction makes it particularly suitable for applications such as stirring bars, stopcocks, bearings etc. Its poor thermal conductivity can be an advantage in some applications but is a problem in apparatus such as digestion vessels which have to be heated. By compression and sintering, containers can be manufactured with high chemical stability for the containment of highly reactive materials.

The inorganic contamination risk arising from the use of PTFE is very low but, in common with all polymers, it exudes low molecular weight compounds from fresh material. Heating to 300 °C before use can help to overcome this problem.

3.4.1.10 Poly(vinylidene fluoride) (PVDF)

$$—[—CF_2—CHF—]_n—$$

This polymer has the lowest melting point of any common fluoropolymer, with a temperature limit of 150 °C. The material has excellent resistance to most chemicals.

3.4.1.11 Poly(chlorotrifluoroethylene) (PCTFE)

$$\text{-[-CClF-}CF_2\text{-]}_n\text{-}$$

PCTFE is the oldest of the fluoropolymers. It has excellent cryogenic properties but is inferior to the other fluoropolymers at elevated temperatures owing to its low melting point (211 °C). Thermally induced crystallization leads to brittleness. In film form it has the lowest moisture permeability. It is normally resistant to corrosive or oxidizing inorganics but is attacked by many organic compounds.

3.4.1.12 Fluorinated ethylene propylene (FEP)

$$—[—(CF_2CF_2)_mCF_2\overset{\displaystyle F}{\underset{\displaystyle CF_3}{C}}—]_n—$$

The chemical resistance of FEP is slightly inferior to that of PTFE, the vulnerable aspect of the structure being the tertiary fluorine. Its dimensional stability is much less affected by temperature changes than PTFE but its upper temperature limit is lower, at 200 °C. The material is translucent and does not absorb in the ultraviolet or visible parts of the spectrum. It is a flexible, high density polymer which resists all known chemicals except molten alkali metals, fluorine and some fluorine precursors at elevated temperatures. The vulnerable tertiary fluorine group is attacked by perchloric acid, and FEP is not therefore suitable for use with this acid. The manufacturing process has been reported to result in the presence of fine metal-rich (iron, zinc, aluminium, copper and manganese) particles embedded in the walls of some FEP bottles.

3.4.1.13 Ethylene–chlorotrifluoroethylene (E–CTFE)

E-CTFE is a 1:1 copolymer of ethylene and chlorotrifluoroethylene.

$$—[—CH_2—CH_2—CF_2\underset{\underset{Cl}{|}}{C}F—]_n—$$

It has excellent resistance to mineral acids and alkalis and most organic solvents. It is used in the construction of tubing, beakers and bottles which can be used at temperatures ranging from cryogenic up to 170 °C.

3.4.1.14 Ethylene–tetrafluoroethylene (ETFE)

ETFE is a copolymer of ethylene and tetrafluoroethylene containing 80% of TFE by weight.

$$—[—(CH_2—CH_2)_m—CF_2—CF_2—]_n—$$

It has better mechanical properties than FEP and, being easier to mould, is generally purer than FEP. Its stiffness is similar to that of PVDF and it is lighter than many of the other fluorinated polymers such as PTFE, PFA and FEP. As with the other fluorinated polymers it exhibits excellent resistance to mineral acids and alkalis and most organic solvents. It is used in the construction of beakers and bottles, which can be used at temperatures up to 180 °C and will normally withstand 200 °C over long periods.

3.4.1.15 Perfluoroalkoxy fluorocarbons (PFA)

These are translucent thermoplastic copolymers of tetrafluoroethylene and a perfluoroalkyl vinyl ether, which can be worked without special procedures. They have a similar backbone to PTFE but with perfluoroalkoxy side chains. PFA does not absorb in the ultraviolet or visible part of the spectrum and has a melting point of approximately 320 °C.

$$—[(CF_2—CF_2)_m—(CF_2—\underset{\underset{\underset{\underset{C_nF_{2n+1}}{|}}{O}}{|}}{C}F)]_p—$$

The chemical resistance of the polymer is similar to that of PTFE but it covers a greater temperature range (– 270 to +260 °C). PFA is particulary resistant to absorption even at 200 °C making it suitable for the storage of trace metal solutions and sample digests. It is relatively insensitive to dimensional changes due to temperature.

3.4.1.16 Polyamide resins

$$H—[NH(CH_2)_6—NH—CO—(CH_2)_4—CO]_n—OH$$

The nylons, such as nylon 6,6, are a series of thermoplastic linear polymers having repeating amide linkages along the backbone. They offer only moderate chemical resistance, being attacked by strong mineral acids, oxidizing agents, phenols, some chlorinated hydrocarbons and some salts.

3.4.1.17 Acrylonitrile–butadiene–styrene (ABS)

This is a family of thermoplastic polymers containing at least 13% of acrylonitrile, 5% of butadiene and 15% of styrene. They are completely resistant to most aqueous acids, alkalis and solvents but disintegrate in concentrated oxidizing acids such as nitric and sulphuric acids. They undergo cracking when in contact with certain chemicals and under stress.

3.4.1.18 Poly(ethersulphone) (PES)

$$—[C_6H_4—\underset{\underset{O}{\|}}{\overset{\overset{O}{\|}}{S}}—C_6H_4—O—]_n—$$

PES is a clear, strong and tough straw-coloured polymer. It is less subject to hydrolytic attack than polycarbonate. It is resistant to acid, bases, aqueous solutions, aliphatic hydrocarbons and alcohols but is dissolved by dimethyl sulphoxide, aromatic amines, nitrobenzene and certain chlorinated solvents.

3.4.1.19 Poly(ethylene terephthalate)(PET)

$$—[CO—C_6H_4COO—CH_2—CH_2—O]_n—$$

PET is a versatile plastic which is widely used in the food industry. It has a melting point of 254 °C and it can be converted into either amorphous or semicrystalline products; when maintained at above 80 °C for a period of time it will crystallize spontaneously. Its resistance to chemicals and solvents is greatly enhanced as the crystallinity of the polymer increases. Possible contaminants are titanium, germanium and antimony.

3.4.1.20 Poly(etheretherketone) (PEEK)

$$—[C_6H_4—CO—C_6H_4—O—C_6H_4—O]_n—$$

PEEK is one of the more specialized plastics which is finding increasing application in the field of chromatography. It is a melt processible thermoplastic which at room temperature is tough, strong and rigid. It has excellent chemical resistance to a wide range of organic and inorganic chemicals. It will, however, absorb 15—20% of methylene chloride and is dissolved by concentrated sulphuric acid.

3.4.2 The Trace Element Content of Plastics

Many trace elements can become incorporated in the polymers which are used in the construction of laboratory apparatus. This can occur both during the manufacture of the polymer feedstock and during processing. The metal concentrations which are found in laboratory apparatus and containers vary between suppliers but some overall trends can be noted from published data (Table 3.3).

Table 3.3 The trace element content of some plastics (ng/g) (Moody and Lindstrom, 1977)

	LDPE	HDPE	PP	PS	PC	PVC	TFE	FEP	ETFE
Na	1300	15000	4800	2200	2700	20000	160	400	600
Al	500	30000	55000	500	3000		230	200	
Cl	7000	30000	180000		50000	major		800	10^6
K	>5000	>600						90000	1100
Ca		800000							
Ti		5000	60000	1000					
Mn		10	20	20				60	
Co			40		6				
Zn		520000							
Br	>20	800	>5	>1	29000	>6	>2		240
Sn						2.4×10^6			
Sb	5	200	600						
La				0.3			0.6		1
W								700	
Au			0.1	0.04	0.03		0.4		0.4

Table 3.4 shows the concentrations of trace elements leached from plastic containers during one week in nitric acid (1+1).

3.4.3 Permeability of Plastics

An important factor in the choice of a polymer is its permeability (Table 3.5). For

Table 3.4 Metal impurities leached from 500 ml or 1 litre plastic containers in one week by nitric acid (1+1). Concentrations in ng per square centimetre of surface. Leaching experiments were carried out at room temperature except for the Teflon bottles, which were leached at 80°C (Moody and Lindstrom, 1977). nd = not detected

Element	Teflon FEP	HDPE	LDPE	PC
Pb	2	2	0.7	0.3
Tl	≤1	≤1	1	≤0.8
Ba	4	≤0.2	2	0.3
Te	0.6	0.2	≤0.5	0.3
Sn	1	1	≤0.8	0.2
Cd	0.4	0.2	0.2	0.3
Ag	≤8	0.2	nd	nd
Sr	0.2	1	0.2	≤0.2
Se	0.2	0.4	3	0.5
Zn	4	8	2	0.8
Cu	2	0.4	2	0.8
Ni	2	1.6	0.5	0.7
Fe	20	3	3	3
Cr	0.8	0.2	0.8	0.3
Ca	80	0.6	10	3
K	2	2	2	2
Mg	8	0.6	0.7	2
Al	6	1	1	5
Na	6	10	8	3
TOTAL	148	50	38	23

most applications a low permeability is desirable, but for the construction of diffusion tubes higher permeabilities may be an advantage. Permeability is derived from the migration of molecules through microscopic voids between the polymer chains. Anything which facilitates an increase in the size of the voids between the polymer chains and the rate of migration will therefore increase permeability.

Table 3.5 Permeabilities of polymers at 20–30 °C (Levovits, 1966). Units: cm^3 (STP)/cm^2/mm/sec/(cmHg) × 10^{10}

Polymer	Gases			Water
	N_2	O_2	CO_2	
FEP	21.5	59	17	500
PCTFE	0.09–1.3	0.25–5.4	0.48–12.5	3–360
PP	4.4	23	92	700
LDPE	20	59	280	2100
HDPE	3.3	11	43	120
PVC	0.4–1.7	1.2–6	10.2–37	2600–6300
PC	3	20	85	7000
Silicone rubber	1000–6000	6000–30000	106000	
Polyamide	0.1–0.7	0.38	1.6	700–17000

Particularly with the linear unsaturated polymers, increasing the temperature increases the mobility of the polymer chain and hence increases permeability.

The absorption of material by polymers is at least in part due to its migration through the polymer structure. Discoloration of a plastic container is often associated with the migration of material into the polymer matrix, and the removal of such material may be exceptionally difficult.

3.5 SILICONES

Silicone is found in three main applications in the laboratory, as a natural rubber substitute, in greases and as anti-foaming agent.

Silicone rubber provides a purer alternative to natural rubbers for use in the laboratory. Unlike natural rubber, it contains no plasticizers or sulphur and has excellent thermal properties, remaining essentially unchanged from −70 to 260 °C. It has many laboratory applications, most notably as connecting and pump tubing, which exploits its superior elasticity. There are reports of contamination from silicone rubber components when they contain fillers or colouring agents; the large red rubber bungs commonly used in laboratories, for example, may contain large amounts of leachable zinc.

Silicone greases have largely replaced hydrocarbon greases in the laboratory. They are intrinsically purer but contaminated glassware may require soaking in concentrated sulphuric acid to remove the grease.

As anti-foaming agents the silicones offer purity and volatility advantages over octanol, which is commonly employed for this purpose. Their removal may however be difficult.

3.6 METALS

3.6.1 Platinum

Platinum is one of the few highly inert materials which can be employed at very high temperatures. It has a melting point of 1772 °C and a high thermal conductivity, making it an ideal material for the construction of vessels used in ashing procedures in which temperatures of up to 1400 °C are necessary. It does not oxidize in air at any temperature and is highly resistant to attack. It can be used with single mineral acids and fusion materials such as alkali halides and sulphates but is corroded by halogens, cyanides and sulphur. It is attacked by mixtures of nitric and hydrochloric acid and several fusion agents including sodium and potassium hydroxides, oxides and nitrates and molten metals. Platinum is rapidly embrittled if it is heated with phosphorus, arsenic, antimony, selenium, tellurium, lead or zinc. Molten silver, gold and most base metals dissolve platinum. At temperatures above 1000 °C it is also attacked by iron and lead oxides and by silica and silicate under reducing conditions. It has a coefficient of expansion which is almost equal to that of soda–lime–silica glass and is therefore well suited to the construction of sealed electrodes.

The material which is most commonly used for the construction of laboratory apparatus is frequently alloyed with iridium and/or rhodium, as pure platinum is very soft. A 10% rhodium–platinum alloy is stronger than pure platinum but is less resistant to fusion materials. The addition of small amounts of zirconium to platinum or its 5% gold–platinum alloy gives a grain stabilized alloy with improved strength when hot and resistance to attack by base metals.

3.6.2 Stainless Steel

Stainless steel is widely employed in the construction of laboratory apparatus. The corrosion resistance of the alloy depends upon its formulation. The 300 series stainless steels consist primarily of iron and, if the surface preparation has not been properly carried out, or the passivity of the surface has been disturbed, then rusting will occur. The austenitic stainless steel group includes the common alloys 304, 316, 321 and 347. Alloys 321 (Ti stabilized) and 347 (Ta and Nb stabilized) are less widely available. The 316 alloy (16% Cr, 10% Ni) is generally more resistant than 304 (18% Cr, 8% Ni), which is attacked by most mineral acids and halide salts.

3.6.3 Aluminium

Pure aluminium is a soft silvery-white metal which melts at 660 °C. The surface is covered by a protective oxide layer when exposed to air. It is light, easily worked and available as a number of alloys. In the laboratory aluminium and its alloys are commonly encountered in the construction of laboratory apparatus, gas bottles and as aluminium foil. Aluminium is unsuitable for most inorganic trace analyses as it is readily attacked by acids, bases and mercury. Anodization of aluminium increases its resistance over the pH range 4.5–8.5 by the formation of a stabilized protective layer of aluminium oxide on its surface. Containers having foil liners in the caps are not generally recommended and aluminium foil should not be used to cover solutions. Corroded aluminium generates a fine oxide dust which is not only a potential contaminant but also an excellent adsorbent.

3.6.4 Titanium

Pure titanium is a lustrous white low density metal which is easily fabricated and has excellent corrosion resistance. It also has a high melting point (1660 °C). It can be made to burn in both air and, unusually, nitrogen. It is resistant to dilute sulphuric and hydrochloric acids, most organic acids, moist chlorine gas and chloride solutions. Alloys of titanium are used when lightweight strength and ability to withstand high temperatures are important. Titanium is as strong as steel but 45% lighter, 60% heavier than aluminium but twice as strong. It is particularly useful in chloride environments and has found widespread application in the construction of marine sampling and measurement apparatus.

3.6.5 Tungsten

Tungsten has the highest melting point (3410 °C) and lowest vapour pressure of all metals. Very pure tungsten can be worked easily but the impure metal is brittle. The metal oxidizes in air and must be protected from the atmosphere at high temperatures. Its thermal expansion is similar to that of borosilicate glass, making it suitable for the construction of glass-to-metal seals.

REFERENCES

Adams, P.B. (1972) in *Ultrapurity: Methods and Techniques,* M. Zief and R.M. Speights (Eds.), Dekker, New York.

Hetherington, G., and Bell, L.W. (1972) in *Ultrapurity: Methods and Techniques*, M. Zief and R.M. Speights (Eds.), Dekker, New York.

Levovits, A. (1966) *Mod. Plast.* **43**, 139.

Moody, J.R. and Lindstrom, R.M. (1977) *Anal. Chem.* **49**, 2264–2267.

Robertson, D.E. (1968) *Anal. Chem.* **40**, 1067–1072.

4 Storage

4.1 INTRODUCTION

In an ideal world all analyses would be carried out in situ, samples would not have to be collected and sample preparation would be unnecessary. Unfortunately this is rarely the case and in most procedures samples have to be collected, placed in a container, transported to the laboratory and then subjected to an extensive sample preparation procedure. All this takes time, and the real life analysis is often a conflict between the practicalities of carrying out a large number of analyses and the deterioration of the sample which occurs during the intervening time. The requirements for sample stability are at their most extreme in the archiving of samples to be analysed in future years. There is much which can be done to improve the chances of the sample remaining unaltered during these steps. From an understanding of the mechanisms which govern the stability of material in storage it is possible to predict conditions which should favour the prolonged life of analytical samples and their solutions.

4.2 FACTORS INFLUENCING STABILITY

Many factors influence the stability of a sample and the analyte, but certain general guidelines are useful in the development of new methods. Although each individual analyte and sample matrix will suffer from its own peculiar problems, when losses occur there are underlying chemical, physical or biological reasons. Stability predictions can be highly successful for standard solutions, but the varying and complex nature of samples may make the results with real samples less reliable. It is rarely possible to avoid storage completely but, as most problems increase with time, every effort should be made to carry out the analysis as soon as possible after obtaining the sample or preparing a solution of it for analysis.

> No storage protocol is sufficiently effective to preserve the sample integrity for ever. Sample deterioration starts at the point of collection, all we can hope for is to slow it down

There are a number of general mechanisms which underlie the problems arising from the storage of dilute solutions. These can be broken down into factors which have chemical, physical or biological origins.

4.2.1 Chemical Factors

The role of a container in maintaining the integrity of a solution has a number of aspects. Firstly, the container should not contaminate the solution. Even at 18 °C water can leach up to 45 mg of sodium out of 1 m^2 of glass in 7 days; at higher temperatures this level will be increased. At the same time as the sodium is coming into solution other glass components will also be leached, resulting in contamination from silicon, boron, calcium etc.

A sample component in solution is normally only one side of an equilibrium between the analyte which is in true solution and a proportion of it which is attached to the container wall as a precipitate, by adsorption or by ion-exchange. The goal of storage is to maximize the proportion of the analyte which is present in solution. Any process which increases the affinity of the analyte for the container wall will lead to losses from solution. The loss of metal ions from solution is normally enhanced by high pH conditions which result in the formation of insoluble species. The provision of a rough container surface favours the nucleation of a precipitate. The rougher the container surface the larger the surface area and the larger the number of available sorption sites.

The monovalent cations, such as sodium and potassium, which have soluble hydroxides are generally not subject to losses from solution by precipitation. Most polyvalent metals on the other hand are particularly difficult to maintain in solution at near neutral pH owing to hydrolysis and disappear very rapidly from solution (Figure 4.1).

At low pH, however, most metal solutions are stable. For the amphoteric elements stability in solution may also be high under alkaline conditions. The problem

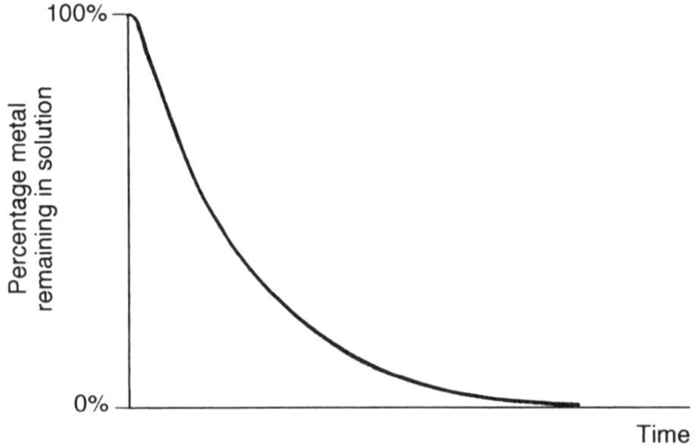

Figure 4.1 Loss of metal from aqueous solution with time

which arises at near neutral pH is largely due to hydrolysis leading to the production of colloids and precipitates of hydrous metal oxides or hydroxides.

$$M^{n+} + nH_2O \rightleftharpoons M(OH)_n + nH^+$$

Other changes in chemical form, such as the disproportionation of iodine in solution, may cause difficulties in speciation analyses.

$$I_2 + H_2O \rightleftharpoons HI + HOI$$

4.2.1.1 Surface activity

The best characterized example of the surface activity of a container material is that of glass. The glass surface can be thought of as consisting of a silicate lattice bounded by a surface of SiOH groups. These are readily ionized, producing an effective ion-exchange surface.

$$SiOH + M^{n+} \rightleftharpoons SiOM^{(n-1)+} + H^+$$

With total element analyses, losses of analyte from aqueous solution occur for a number of reasons, but in most cases these are associated with the formation of neutral species. A useful general approach to the stabilization of the analyte in aqueous solution is therefore to ensure that it is present in a charged ionic form. In the case of metals, the main problem reaction which leads to the loss of metal from solution is, as discussed previously, the hydrolysis reaction

$$M^{n+} + nOH^- \rightleftharpoons M(OH)_n$$

To minimize the formation of the insoluble product, all that is needed is to limit the concentration of OH^- ions in solution, i.e. to acidify the sample.

The acidification of the sample has an important secondary stabilization effect. It has been mentioned previously that a glass surface can be considered as a silicon lattice with terminal acidic OH groups which can act as an effective ion-exchange surface.

$$SiOH + M^+ \rightleftharpoons SiOM + H^+$$

Increasing the acidity of the solution results in such reactions being driven to the left, protonating the glass surface and releasing the metal back into solution.

There are a variety of ways to achieve a similar effect, any reaction leading to the metal ion being maintained in solution as an ionic species *may* be a suitable means of stabilizing the sample.

High salt content:

$$Cd^{2+} + 4Cl^- \rightleftharpoons CdCl_4^{2-}$$

Adding a chelating agent:

$$Pb^{2+} + EDTA^{4-} \rightleftharpoons [PbEDTA]^{2-}$$

For the amphoteric metals, stabilization can even be achieved under alkaline conditions:

$$Zn^{2+} + 4OH^{-} \rightleftharpoons Zn(OH)_4^{2-}$$

An alternative approach to ensuring that species remain in solution is to block off reactive sites on the container wall. With glass containers this can be carried out by reacting the surface Si–OH groups with a suitable reagent such as trimethylchlorosilane. The 'silanization' of the glass results in the production of a hydrophobic surface lacking the ion- exchange sites of the native material.

$$Si{-}OH + ClSi(CH_3)_3 \longrightarrow Si{-}O{-}Si(CH_3)_3 + HCl$$

The reaction is catalysed by the presence of some dimethyldichlorosilane.

4.2.1.2 Oxidation

Oxygen, which makes up 23.2% of the atmosphere, is the most common oxidant to affect the stability of solutions. Its solubility in water is very high and, at 20 °C, 9.1 mg of oxygen can be present in a litre of water. Oxygen is a fairly strong oxidant with a redox potential for the reaction

$$O_2 + 4H^+ + 4e^- \rightleftharpoons 2H_2O$$

of 1.23 V at 25 °C. In neutral pH water, the redox potential is still 0.82 V, sufficient to oxidize many species (Table 4.1).

Table 4.1 Redox potentials of some solution species

Reduced species	Fe^{2+}	I^-	SO_3^{2-}	Sn^{2+}	Cr^{2+}
Oxidized species	Fe^{3+}	I_2	SO_4^{2-}	Sn^{4+}	Cr^{3+}
E^o (V)	0.77	0.53	0.17	0.15	-0.41

4.2.1.3 Atmospheric carbon dioxide

The concentration of carbon dioxide in the atmosphere is variable, but generally lies within the range 0.02–1% by volume. Carbon dioxide is readily soluble in water, giving a solution containing approximately its own volume of carbon dioxide (0.2% by weight). The dissolution of atmospheric carbon dioxide results in a significant lowering of pH and as a consequence distilled water is normally acidic. Carbon dioxide will also react with alkali to give an insoluble carbonate and precipitates other insoluble metal carbonate species. It reacts with sodium thiosulphate to give $NaHSO_3$.

4.2.1.4 Photochemistry

Some chemicals are particularly susceptible to photochemical decomposition by natural light. The storage of chemicals which have a known susceptibility to photochemical degradation should be treated with particular care and such compounds should be stored in the dark. Brown glass is not generally recommended in trace analysis because of the impurities the glass may introduce into the sample. In some cases, such as with iodide and oxalic acid, photochemical decomposition is well established from classical procedures and some of the preservation procedures developed for high level analyses may be useful in stabilizing lower concentrations. The photochemical stability of metal complexes, such as the copper, lead and mercury diethyldithiocarbamates, may be impaired by the instability of the ligand.

4.2.1.5 Gas adsorption

Losses and contamination from container walls are not restricted to solution studies. A more difficult situation can arise in the storage of samples for trace gas analysis. Most materials have highly reactive surfaces and losses of small quantities of gas by adsorption are inevitable. Losses onto the metal oxide coatings which cover most metal surfaces may make the storage of such samples impossible.

4.2.2 Physical Factors

4.2.2.1 Sorption onto container walls

Container walls are never completely inert and they can play an important role in the loss of analyte from solution during storage. The role of the container as a nucleation surface for the precipitation of solid material from solution has been discussed in the previous section. Although nucleation of the precipitate would normally occur within the bulk of the solution onto dust particles or other extraneous material, these may not be present in high purity reagents and the deposition then occurs on the container wall.

4.2.2.2 Permeable containers

The materials which are employed in the construction of containers vary significantly in their permeability. Permeation through a container wall can result in a number of analytical problems ranging from sample contamination through to loss.

The ingress of material into a container is most frequently associated with gas permeation. With samples which have been taken for the analysis of a particular chemical species, the introduction of oxygen into the sample through the container wall can result in unwanted speciation changes. Even in a total metal analysis a speciation change can be troublesome. Take for example the analysis of a neutral water sample for dissolved iron in which the iron is largely present in its iron(II) state. Oxidation of the iron to iron(III) will result in the generation of the insoluble hydrous iron(III) oxide, which can precipitate or sorb onto the container surface, resulting in analyte loss from solution.

The movement of material out of the container by permeation takes two main forms. The long term storage of solutions in unsuitable containers is known to result in the loss of water through permeation. The resulting change in concentration of the analyte may go unnoticed unless the precaution has been taken to weigh the sample before storage. The second complication arises with analytes which are in equilibrium with volatile species such as ammonium and sulphide salts. Although only a small proportion of ammonia or hydrogen sulphide may be present out of solution, the loss of the gas through the container wall will lead to the re-establishment of the equilibrium, generating further gas to permeate out of the container.

4.2.3 Biological Factors

The biologically derived aspects of sample stability can arise both actively, as a result of live organisms, and through the presence of inactive cellular components.

4.2.3.1 Active biological effects

Active biological processes result in the turnover of sample material as it is used as an energy and nutrient source. The nature of the organism which is involved will depend upon the type and history of the sample, but it will frequently be bacterial or algal. The most obvious consequences are that the sample itself is consumed, resulting in a mass loss, and that nutrient materials are taken from solution. Major speciation changes can occur not only as a result of the metabolism of the species of interest but also because of changes in the chemical nature of the sample matrix as a result of oxygen depletion or acidification of the sample.

Many of the commonly employed procedures for the stabilization of chemically derived sample deterioration are also effective in limiting biological activity. The addition of poisons such as azides, mercury salts, carbon tetrachloride and antibiotics may in some cases be appropriate. Acidification of the sample or freezing will often achieve the same goal but without the addition of reagents, some of which are difficult to purify.

4.2.3.2 Passive biological effects

It is unfortunately possible for biological organisms to influence a chemical analysis even when they are dead. There are two main routes by which this can occur. The rupture of cells results in the release of cellular fragments which can be an active adsorption surface. In addition, enzymes are not inactivated by the cellular disruption and can continue to affect the chemical composition of the sample. Although enzyme poisoning and freezing are the most effective means of limiting such changes, it should be noted that these preservative methods may themselves result in the lysing of cells due to osmotic pressure changes, chemical decomposition or freeze fracturing. Liquid samples should normally be filtered immediately after collection and before the addition of additives or freezing.

4.3 CONTAINERS

4.3.1 Design

The design of a vessel can influence its suitability in a number of ways. Containers frequently incorporate a number of different materials, each of which can influence the suitability in a specific application. Each analysis will inevitably require particular characteristics of the container. In some circumstances it may only be necessary to prevent significant liquid leakage during storage. In other cases more efficient sealing may be necessary to prevent gas phase transfer of the analyte out of the container.

Freezing is an effective storage method for many analytical procedures, but care must be taken to ensure that differential contraction of the container and its cap does not result in the cap becoming partially released during storage. Freezing full bottles of aqueous solutions will normally result in the bottle fracturing owing to the expansion of the sample during freezing.

It should not be assumed that containers are made of a single material. A bottle which is advertised as being constructed from polyethylene may well be fitted with a polypropylene cap. This cap may in turn have a metal-surfaced card liner. Whereas a solution stored upright in such a bottle will spend most of its time in contact with only the polyethylene, movement of the bottle and solution reflux inside the container will ensure that sufficient contact with the cap material occurs to cause significant contamination. In some cases it may be possible to adapt such bottles by replacing the cap liner with, for example, sheet PTFE, but careful initial selection is preferable. It is also normal practice for cap materials to be strengthened or coloured by inorganic additives. A cap which is described as being made of polypropylene may contain significant quantities of potentially contaminating filler material.

When solutions are poured from a screw-capped conventional container, some contact between the solution and the threaded portion of the container will be inevitable. The threaded area is particularly susceptible to the build-up of contaminating material from the cap and seeping material. In seawater analyses, deposits of salt crystals are frequently found around the thread and are washed into the sample when it is poured from the bottle. One approach to the problem of thread contamination is the use of containers having inverted caps (Figure 4.2).

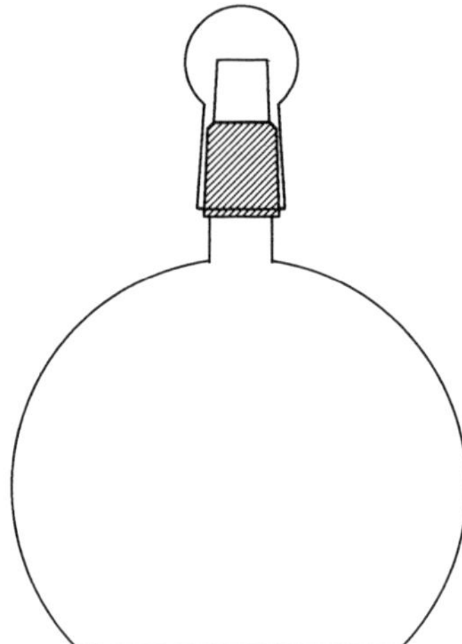

Figure 4.2 A container with an inverted cap based on standard components

These reduce contamination from the threaded cap area of the bottle but are unfortunately difficult and expensive to obtain. They may therefore not be practicable alternatives for the storage of large numbers of samples.

4.3.2 Cleaning and Pretreatment

For trace analysis applications it is extremely rare to be able to assume that new or conventionally cleaned apparatus is suitable. Almost without exception it will be necessary to carry out a rigorous cleaning procedure when a new container is first put into use. The extent and nature of subsequent cleaning depends very much on the application in which the container is to be used.

> Used apparatus may be cleaner than new stock

Each analysis will require the development of an individual cleaning procedure which must be designed to take into account the potential for analyte loss and contamination. Whilst contamination from dirty apparatus is predictable, more subtle effects can occur, such as those engendered by the permeation of volatile materials into and out of the container during storage.

Having selected the type of vessel to be employed, the first step is to ensure that the available vessels are both consistent and that their condition can be returned to a known fixed state between analytical applications. Consistency is a prerequisite of carrying out any blank analysis and correction. The following guidelines will assist in attaining consistency.

(1) Use a batch of apparatus with a known and common history

> Choose apparatus from a single batch of known origin to ensure consistency

Although this is most readily achieved if the apparatus comes from a single new batch, expense may make this impracticable. The following types of previously used flasks should not be mixed with new containers:

(a) flasks which have been used to make up or store high concentration metal standards;

(b) previously silanized bottles;

(c) glass containers which have contained caustic solutions. The surface etching effect results in a surface radically different from that of a new container.

(2) Select perfect containers

It is not uncommon for minor imperfections to be present on the (internal) surface of a container. Experience with such containers is that they should be discarded at an early stage, as they will frequently either contaminate the sample or provide a suitable surface on which analyte losses can occur.

> Flawed containers contribute to both contamination and analyte loss; discard them

Such a container, if inadvertently used for a blank measurement, could adversely affect the results of many analyses.

(3) Reveal the true nature of the container surface

Most new containers are coated to a greater or lesser extent with grease or other production materials. This may block active sites on the container surface initially, but its ability to do so throughout the lifetime of the container is not guaranteed, nor is the coating likely to be contaminant free or evenly applied.

Only a naked container surface reveals its true identity

Degreasing can normally be achieved using commercial detergents. If necessary, the less toxic and environmentally less damaging chlorinated solvents, or more rigorous oxidation procedures, can be employed. Compatibility of the cleaning materials with both the final analysis and the chemical nature of the container to be cleaned must be considered at this early stage. Phosphate-based detergents, for example, are not ideal for the degreasing of apparatus to be used for phosphate analysis, and chromic acid[†] attacks most plastics and can leave exceptionally high levels of chromium on the apparatus. An extended soak (up to a week at room temperature) in a high efficiency laboratory cleaning detergent is normally adequate. The time scale can be shortened by using a heated detergent bath. All traces of the detergent should be removed by thorough and careful rinsing with good laboratory grade purified water, as the detergent itself can contain contaminants. As these contaminants are largely in solution, rinsing should remove them sufficiently if further treatment is envisaged.

Once production-derived impurities have been removed from the container surface, more specific cleaning procedures can be carried out. One such procedure for the generalized minimization of trace metal impurities has been described by Moody and Lindstrom (1977). This employs a sequential leaching of the container with hydrochloric and nitric acids, which were shown to provide complementary cleaning actions.

(1) Fill the container with hydrochloric acid (1+1, analytical reagent grade) and allow to stand for one week at room temperature (80 °C for PTFE).

(2) Empty, rinse with distilled water, refill with nitric acid[††] (1+1, analytical reagent grade) and allow to stand for one week at room temperature (80 °C for PTFE).

(3) Empty and rinse with distilled water. Refill with the purest available distilled water.

(4) Allow the container to stand, changing the water periodically to maintain the cleaning action.

---

† Chromic acid cleaning mixture is prepared by adding 100 ml of concentrated sulphuric acid slowly, and with constant cooling, to a solution of 5 g of sodium dichromate in 5 ml of water. Protective clothing must be worn.

†† N.B. Nitric acid is an oxidizing agent and, particularly at elevated temperatures, will attack the polymer side chains, resulting in acidic surface groups. Careful consideration should be given to whether this acid is suitable for the specific application.

(5) Rinse with the purest water and allow to dry in a particle- and fume-free environment.

If only limited quantities of apparatus have to be cleaned, steam cleaning with nitric or hydrochloric acid (Figure 4.3) can be effective.

It should be noted that the extended treatment of plastics with acids can lead to the permeation of acid into the polymer, resulting in the release of acid into the samples during use. Acid-washed glassware can also change the pH of unbuffered solutions.

4.3.3 Transfers between Vessels

The quantitative transfer of material is of course just as valid in trace analysis as it is at higher levels. At trace levels the physical transfer of a liquid may not be sufficiently stringent to ensure that all the analyte is transferred. This is particularly the case if the container surface is active and the analyte is not completely in solution. In order to avoid losses, every effort should be made to restrict the number of transfers to the absolute minimum. This will ensure that losses from analyte

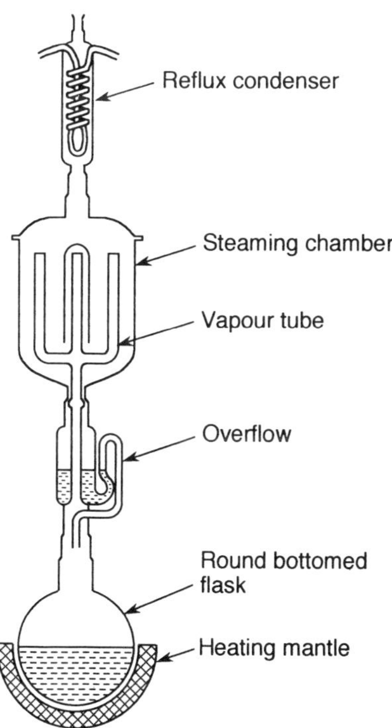

Figure 4.3 Apparatus for steam cleaning with acid vapour. (Adapted from Tschopel *et al.*, 1980, by permission of Springer-Verlag, Heidelberg

remaining attached to the container surface are minimized and that contamination is reduced.

> **Minimize the number of transfers to reduce analyte losses and contamination**

If transfers between containers are necessary, steps must be taken to ensure that surface-bound analyte is known to be in solution prior to the transfer. In most cases it is necessary to ensure that the analyte is brought back into the bulk of the solution before to the determination step. Under most circumstances steps will have been taken to maintain the analyte in solution during temporary or long-term storage and, if this has been completely effective, all analyte should be removable from the container without trouble. Care should however be taken to ensure that the composition of the solution is not significantly altered during the washing stages. If an element which is particularly sensitive to hydrolysis has to be stored under highly acidic conditions, washing the remainder of the sample from the container with distilled water may significantly alter the solution pH, resulting in deposition of the analyte from solution. Under such circumstances rinsing should be carried out using a solution having the required pH to maintain the analyte in solution.

Some analytical procedures have been developed which are known to employ preservation methods which, from the point of view of maintaining the analyte in solution, are not ideal. The preservation of seawater samples by freezing is not uncommon. It is often the only method available for speciation analyses as the addition of reagents frequently alters speciation. During the freezing of seawater, however, ice formation and changes in solution composition result in the deposition of a sediment. The analysis of seawater which has been stored in this manner may necessitate allowing the sample to re-equilibrate after thawing to bring all the analyte back into solution. In the case of aluminium and antimony analyses, this may require the thawed sample to be held at room temperature for 24 hours prior to the analysis.

4.4 THE STORAGE OF SOLUTIONS

In a large proportion of analytical procedures both standards and samples are present in solution. In a trace analysis these are generally extremely dilute solutions containing quantities of analyte so small that only very small absolute losses or amounts of contamination are necessary significantly to influence results. Indeed, the surface area of the container can be sufficiently large for monolayer coverage with analyte to result in complete loss from solution. It has long been recognized

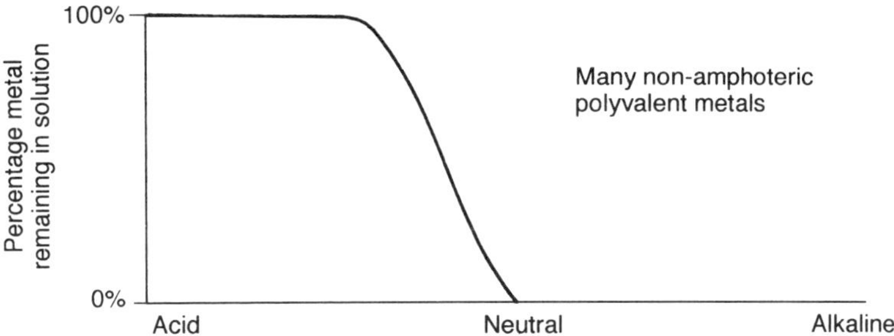

Figure 4.4 The loss of dissolved metals during storage

that certain analytes cannot be stored for significant time periods without measures being taken to stabilize them. The behaviour of solutions of the main group and transition elements provides a good example of the problems which can be encountered if preventative measures are not taken. If trace metal standards are prepared under various pH conditions and stored, the following day little of the metal may remain in solution (Figure 4.4).

The situation is slightly different, however, if the element exhibits amphoteric behaviour (Figure 4.5).

As a general guideline, therefore, trace metal solutions are likely to be at their most stable under highly acidic conditions. A pH of less than 2 is often recommended for this purpose.

> Trace metal solutions are generally at their most stable under acid conditions

Care must of course be taken to ensure that the analyte element does not form an insoluble species with the anion of the acid which is used to lower the pH.

4.4.1 Some Examples

An understanding of the basic principles behind the losses of material from solution can be helpful in the design of new experimental procedures. Each analyte has its own reaction chemistry and consequently different requirements for solution storage. An appreciation of this reaction chemistry may lead to more rapid success in method development.

The following examples illustrate aspects which must be considered when designing an analytical procedure. Underlying the choices which must be made is always the chemistry of the processes which are likely to take place during storage.

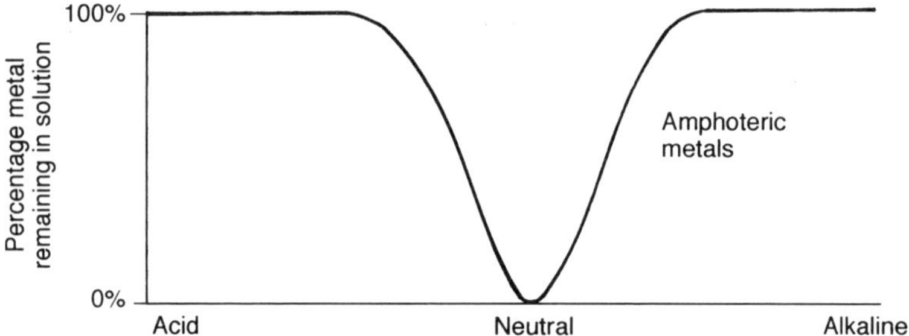

Figure 4.5 Stabiity of dissolved amphoteric elements under various pH conditions

4.4.4.1 Iron standards

The solution chemistry of iron exemplifies the difficulties which can be encountered in the preparation and stability of solutions of trace metals. The problem with the stability of this metal arises from the formation of insoluble hydrous iron(III) oxides at near neutral pH.

$$Fe^{3+} + 3OH^{-} \longrightarrow Fe(OH)_3$$

The behaviour of iron in solution can best be seen from a study of its pE (or E_H)/pH diagram (Figure 4.6).

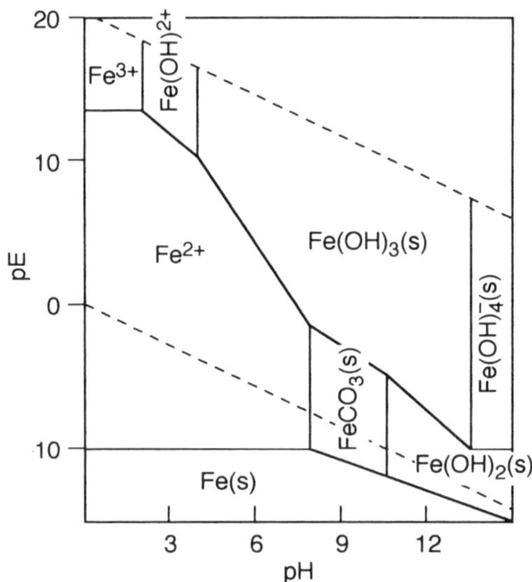

Figure 4.6 The pE/pH diagram for iron species (pE = – $\log_{10}$ {e^-})

With this element there are three main methods to ensure that the iron remains in solution. The most commonly employed approach is to acidify the sample to ensure that hydrolysis does not occur. Alternatively, reduction of the iron to iron(II) under acidic conditions or the addition of a suitable complexant should ensure that charged species are present.

4.4.4.2 Mercury standards

Of the most commonly determined trace metals, mercury is distinctive, as it can be present in the volatile elemental form.

$$Hg^{2+} + 2e^{-} \rightleftharpoons Hg^{0}$$

This is a volatile species which partitions into the gas phase above the sample solution and which is capable of diffusing through the container wall. In order to limit the permeability of the mercury during sample storage, it is therefore necessary to stabilize the solution by adding a suitable oxidant to keep the element in its soluble Hg(II) ionic form. Traditionally either permanganate or dichromate salts are employed. The former leads to some deposition of manganese dioxide and dichromate is often favoured owing to its solution stability. A compatible acid (one which is not itself oxidized by the permanganate or dichromate) can also be added to maintain the pH of the solution below values at which hydrolysis would occur.

4.5 PRESERVATION AND STORAGE OF SOLIDS

It is generally true that the storage of solids presents less of a problem than is the case with solutions. The very nature of a pure solid, consisting of an outer exposed surface enclosing a solid core, affords some degree of protection to the core material. The limitations of diffusion into the centre restrict the movement of oxygen into the core and the possibilities of biological activity are reduced. Similarly, the movement of analyte out of the solid and the direct transfer of contaminants from the container to the sample are difficult. Many samples are not pure solids, however; they are in fact mixtures. Soil for example, until it is dried, is a mixture of soil particles and soil organisms with water. Such a sample will have features in common with both solutions and solids. Not only will it be subject to the many potential contamination and loss problems associated with solutions, but the presence of the water will induce other difficulties with some samples. If the samples are frozen to minimize deterioration or diffusion, freeze fracturing of the particles may occur. Significant sample separation can also occur by the sublimation of water.

The deterioration of solid samples can in many cases be minimized by comparatively straightforward means such as drying or freezing. The choice of container is as important as it is with solutions, but for slightly different reasons. Although the direct surface to surface transfer of contaminants from the container wall is limited, the mechanical abrasion of the container wall can introduce significant quantities of material into the sample. Not only must the container materials be chosen carefully, but consideration must also be given to sample movement during transport.

Drying is normally carried out either above or below ambient temperature to enhance the rate of water loss (see Chapter 7, Working Practices). Conventional ovens can be used for those samples which are not adversely affected by heating in the atmosphere. The drying should normally be carried out slowly, at temperatures below 100 °C. The rate of water loss is significantly enhanced by taking the temperature above the boiling point of water, but this may lead to explosive volatilization of the water, distributing the sample around the oven. Fan assisted ovens may dry material more rapidly, but their potential for contaminating the sample is unacceptably severe. In order to protect more delicate samples from the atmosphere the heating can be carried out under a flowing stream of inert gas or in a vacuum oven. Freeze drying (lyophilization), involving the sublimation of water from a frozen sample under vacuum, is a useful technique for delicate samples. However, no drying technique can guarantee the conservation of speciation.

4.6 GASES

At the trace level, the storage of trace level gas samples for extensive time periods can at best be described as unreliable. Gaseous diffusion and adsorption onto container walls combine to ensure that long term storage of such samples is not generally advisable. Whereas the state of development of methods for the storage of solid and liquid samples is well advanced, there are few corresponding preservative methods which are currently considered to be suitable for the stabilization of gaseous samples.

REFERENCES

Moody, J.R. and Lindstrom, R.M. (1977) *Anal. Chem.* **49**, 2264–2267.

Tschopel, P., Kotz, L., Schulz, W., Veber, G. and Tolg, G. (1980) *Fresenius' Z. Anal. Chem.* **302**, 1.

5 Reagents

5.1 INTRODUCTION

All trace analysis procedures depend for their success on the availability of pure reagents. Although the 'value added' market for speciality chemicals has increased greatly in recent years, it will never be possible for the suppliers to provide a sufficiently wide selection of chemicals to satisfy all the requirements of the analyst. Each procedure has its own criteria of reagent purity and it will therefore be necessary to develop particular purification protocols for each analysis. The performance of the reagent in the procedure can be the only criterion which will determine the success of a purification.

With such an open brief, it is therefore impossible to provide definitive recipes for the purification of all reagents for all applications. This chapter sets out to discuss some of the options which are available for the production of pure reagents. For more specific details on methods which have been employed for the purification of chemicals (but unfortunately rarely for trace analysis applications), the reader is referred to the comprehensive compendium by Perrin and Armarego (1988).

5.2 HAZARDS

The safety aspects of all laboratory work have received increasing attention in recent years, and the move towards regulation has now reached the laboratory in many countries. The appraisal of risks associated with potentially hazardous operations and the adoption of good laboratory practice are, irrespective of legal requirements, fundamental to all scientific investigations.

> Identify hazards, and how to deal with them, before starting the experiment

One of the first things which must be noted is that most analytical procedures involve risks, which originate from a number of sources. The most obvious risk arises from the use of toxic compounds. Such materials do not however normally present a risk to the investigator until they are used. The risk which results from a chemical in a well sealed and designed container is normally almost nil. It is only when that container is opened that a potential risk results. It may therefore be better to concentrate consideration on the risks that result from activities, rather than on the intrinsic toxicity or other potentially hazardous property of the reagent. A chemical which is well handled should present a negligible risk to the investigator. Handled badly, that same chemical can present an extreme risk.

The physical nature of a reagent can play an important role in the hazardous nature of the material. Whilst there may be no chemical difference, finely powdered material may present a much larger potential inhalation risk than pelleted material. An often forgotten aspect of laboratory life is long term exposure to volatile reagents. In most laboratories a large turnover of air occurs due to the fume extraction systems. In areas where fume extraction is poor, potentially dangerous levels of vapours can build up. This is most often encountered in small instrument rooms or store rooms.

Acute chemical toxicity is not of course the only type of hazard which can be present in the laboratory. A large proportion of the potentially dangerous operations involves apparently simple physical operations such as pipetting or the connection of flexible tubing onto condensers, or electrical shocks.

The time to carry out a risk assessment is, as it always has been, before commencing work. While this book can only include a limited discussion of the safety aspects of trace analysis, it should be noted that there are a number of common risks associated with work of this kind. The early and formal identification of hazardous materials and operations is now an integral part of legislation, and will require some adaptation of working practice even by established and experienced chemists, some of whom may consider that such an approach belittles their expertise. Needless to say, an approach which is designed for those with little knowledge and expertise is an invaluable, if sometimes arduous, training aid for all.

5.2.1 Reagent Selection

Information should be obtained on the hazardous properties of all the chemicals used in the procedure before starting work. This will involve the collation of available toxicity data, recommended maximum exposure levels and the potential for hazardous reactions to occur. Data to be collected include occupational exposure levels, LD_{50} values (i.e.the dose that is lethal to 50% of a population), flash-points, carcinogenicity etc. If at this stage reagents are identified as being unacceptably toxic or otherwise hazardous, attempts should be made to reduce the potential risk by substituting less problematical materials.

There can be few analytical procedures which are free from all risks. It is the role of the professional chemist to minimize potentially dangerous aspects of the work which might endanger himself, his fellow workers, his workplace and the environment. If the work is to be carried out by a less experienced worker, this should be taken into account when appraising the risk and suitable training should be given. The recognition of risks is the first step towards minimizing them. In many cases, simply replacing a hazardous chemical by a related one may leave the procedure essentially unaffected but result in significant safety improvements. Consideration of the depletion of the ozone layer by halogenated solvents will also dictate the substitution of certain commonly employed solvents by materials having less environmental impact.

Substitute safer reagents for the hazardous ones found in many standard methods

There are still many procedures in use which were developed before their hazardous aspects had been identified, and it should not be assumed that methods found in the literature necessarily employ the safest of procedures. Our understanding of the toxicological properties of chemicals is continuously under development and new risks in the use of particular procedures are identified every year. In many cases the reagent or procedure poses what would now be considered to be an unnecessarily high risk. Benzene, a popular solvent in the past, is now banned in many establishments because of its carcinogenic activity. This is a particular example where an effective (but not entirely hazard-free) substitute, toluene, is readily available. This substitution is frequently successful and significantly reduces the hazards in the procedure. Typical examples of what can be achieved by suitable reagent substitutions are given in Table 5.1.

Table 5.1 Some reagent substitutions to be considered as options for the reduction of toxicity risks. Occupational exposure standards quoted as 8 hour time weighted average (TWA) values in mg/m3 (data from Health and Safety Executive, 1992)

Reagent		Substitute	
Chemical	8 hour TWA	Chemical	8 hour TWA
CCl_4	12.6	CH_2Cl_2	350
$CHCl_3$	9.8	CH_3CCl_3	1900
C_6H_6	16	$CH_3C_6H_5$	188
CH_3OH	260	C_2H_5OH	1900
n-C_6H_{14}	70	cyclo-C_6H_{12}	340
		n-C_5H_{12}	1800
		n-C_7H_{16}	1600

5.2.2 Handling Chemicals

Having selected the reagents and recognized some of the problems involved in their presence, the handling of the materials should be considered both to minimize exposure and to reduce risks from, for example, spillage or fire. The assessment process should have identified the particular hazards which might be associated with a particular chemical, but under most circumstances an unopened bottle sitting on the shelf is unlikely in itself to present a risk. It is in fact the handling of the chemical and subsequent reactions which it might undergo which generate the risk. Problems may start simply with the opening of an old reagent bottle which has built up an internal gas pressure during storage. Before the opening stage the operator should anticipate situations in which exposure might occur: dust or vapour inhalation and spillage being two of the more obvious examples. Not only should the chemist take steps to prevent exposure to himself by wearing suitable protective clothing, but steps should be taken to ensure that, in the case of a hazardous event, corrective action can be carried out immediately. Locating a fire extinguisher after a fire has broken out, or finding out how to neutralize a major spill after it has occurred, is too late. It must be borne in mind that the risks generated during an experiment might affect others. Major events may result from minor hazards, a common example being the use of diethyl ether in the same room as a source of ignition such as a bunsen burner and, less obviously, a hotplate or hot-air dryer. The risk to the operator can be minimized in a number of ways. Where there is a choice between protective clothing and an engineered solution such as a fume extraction system, the latter should normally be chosen.

There are a number of simple precautions which can be taken to minimize risks; these are hopefully second nature to the professional chemist but are worth repeating.

5.2.2.1 Acids

The most obvious problems if suitable precautions are not taken, which arise from the handling of concentrated mineral acids, are burns to skin. Safeguards against such hazards involve both the use of protective clothing and good handling technique. The following few tips should be followed to minimize the risk of burns:

(1) do not move concentrated acids around the laboratory unnecessarily;

(2) put all reagent bottles and glassware to the rear of the bench;

(3) only bring apparatus and chemicals which are in immediate use to the front of the bench;

(4) ensure that clothing and hair cannot come in contact with apparatus or chemicals; tie hair back if necessary;

(5) wear suitable protective clothing. Note that 'rubber' gloves are not resistant to all chemicals and that, whilst too heavy a grade of gloves can create a risk by being too clumsy, lightweight clothing may not offer sufficient protection. In the context of trace analysis, it is important to consider at the same time the

potential of protective clothing to contaminate the samples. Gloves that have been coated with talc should particularly be avoided;

(6) do not have flexible rubber or plastic tubing on the bench unless it is in use and secured;

(7) take particular care when pipetting concentrated acids, as accidental discharges onto the lower parts of the body can be particularly serious. It is easy to become complacent with acids, as the hands resist burns significantly better than more sensitive parts of the anatomy. Pipette fillers must always be used, but it must be remembered that these are not pipette holders and both the pipette and filler must be held at all times;

(8) wipe up all spillages immediately. You may be aware of the presence and nature of a minor spill but your colleague or a cleaner may not be.

Most of the hazards involved in the handling and use of acids are a fundamental part of the training of any chemist but there are a number of such chemicals which present less obvious risks.

(1) Perchloric acid, one of the most effective acids for the digestion of samples for trace metal analysis, has a tendency to cause explosions. Some organic perchlorates and metal perchlorates are metastable and can (and do) explode with disastrous results. Such explosions and fires have been known to occur during the dismantling of fume hoods which have, at some time during their life, been used with perchloric acid. Employing perchloric acid for digestions, or any other purpose which involves heating, is banned in many laboratories. Those organizations wishing to carry out such digestions require special fume cupboards, with washed walls and a scrubbed exhaust, and must take particular precautions. Transition metal perchlorate salts, which are sometimes used for the preparation of standards, should be treated with equal respect. It is not advisable to use a spatula or similar device to break up a crusted solid.

(2) Hydrofluoric acid causes very severe burns and should only be used by specially trained personnel who have burn treatment preparations immediately to hand.

(3) Chromic acid, frequently used for cleaning glassware, causes severe burns. It should never be stored in sealed containers, especially after use, as the release of carbon dioxide from the oxidation of organic material can lead to a pressure build-up in the bottle.

(4) The combination of oxidizing acids, such as concentrated nitric acid, and organic solvents such as ethanol must be avoided at all costs. Despite the popularity of this combination for the cleaning of glassware by synthetic organic

chemists, the reaction is unpredictably dangerous and there may be a significant time delay before a very violent reaction occurs. It should be noted that organic acids are no exception and can also react violently with oxidizing mineral acids.

5.2.2.2 Organic solvents

A number of specific hazards arise from the use of common organic solvents; here are a few examples.

Diethyl ether Extreme fire hazard, potential explosion hazard if peroxides are allowed to form. The commercial solvent often contains a stabilizer but this will normally be removed by purification. Peroxides should be destroyed before distillation and the distillate should then be stored in the dark under nitrogen. Use the purified material quickly.

Carbon disulphide Extreme fire hazard, toxicity.

Pentane Very volatile at room temperature and fire hazard.

Chloroform and related solvents Toxicity, especially if phosgene is allowed to form by oxidation. A stabilizer is normally added by manufacturers to reagent grade chemicals. This limits phosgene formation but is removed by most purification methods.

Acetic acid This is often thought of in association with the mineral acids and it is forgotten that this is a highly flammable organic 'solvent'.

5.3 THE SELECTION OF REAGENTS

The first thing to establish when considering the purchase of reagents for trace analysis is that the most expensive reagent is not necessarily the best for a particular application. Manufacturers use a variety of terms to describe their products. The more general terms applied to reagents include 'technical grade', 'analytical reagent grade' and 'high purity'. Certain manufacturers have gone to the extent of grouping together to publish specifications of their products as standards, an example of which is the 'AnalaR' standard produced by the then British Drug Houses (BDH) and Hopkin & Williams (BDH, 1984).

> The most expensive reagent is not necessarily the purest for your application, but purification can be an expensive option

The price differences between the reagent grades often reflect the additional costs involved in the analysis of batches, sometimes purification, and storage costs. The final designation of a grade for a batch will depend on the analytical results in relation to specifications, and not necessarily on specialized purification procedures; some batches of reagent are purer than others. If an element of interest is not mentioned on the label it does not mean that it is absent from the reagent. Some manufacturers undertake specialized clean-up procedures and will supply, for example, mineral acids prepared by sub-boiling distillation.

> Screen batch purity, note batch numbers and then try to buy in bulk from the same batch

Batches of reagent vary in the levels and nature of their impurities. One batch of reagent cannot therefore be expected to have the same contaminant level as a different batch; each batch must be checked for its suitability in a particular analysis and in extreme cases reagents may have to be ordered by batch number to ensure uniformity in a large analytical project. In order to preserve the integrity of these reagents they should be reserved for the specific analysis and should not reside on the open laboratory shelf.

Ultimately, the choice of reagent grade will be that which will give an acceptable result at lowest cost, whether this be in financial terms or in the manpower costs involved in the laboratory purification.

> Container selection and preparation govern reagent purity

High purity reagents must be stored carefully to maintain their integrity. This is particularly true for aggressive reagents such as concentrated mineral acids. There is a conflict to be dealt with here, as the requirement to check the purity of all batches of reagent before use implies the purification of large batches. Contamination from storage is minimized, however, by the preparation of small batches for immediate use. The selection of storage containers and their preparation are dealt with in greater detail elsewhere in this book.

5.4 PURIFICATION OF ORGANIC SOLVENTS AND LIQUID REAGENTS

When considering the purification of organic liquids, there is no distinct demarcation line between the organic solvents and reactive organic reagents employed in the procedure. In most cases the only distinction which can be made is that of scale, solvents being generally employed in larger volumes. The large volumes of an organic solvent which are required usually dictate that the chosen method should be suitable for the purification of litre quantities without recourse to plant scale apparatus.

Purity criteria for organic solvents which are to be employed in a trace inorganic analysis are normally somewhat different from those applied in the selection of materials to be used in either trace organic analysis or synthetic chemistry. Whilst there may be examples where more stringent requirements are necessary, the main criterion for purification remains the need to ensure that the solvent is pure in the context of the particular analysis. A solvent to be employed in the extraction of a metal complex from water must firstly, therefore, produce a low reagent blank. Metal impurities will dominate the blank and the presence of organic impurities will normally remain a secondary consideration. There are a number of circumstances, however, when it may be necessary to carry out solvent purifications for reasons other than to reduce the concentration of analyte in the solvent. In colorimetric procedures, for example, there may be coloured impurities in the solvent or there may be residual oxidizing materials in the solvent which bleach the coloured complex to be measured. Although the impurity levels in the solvent may be small, it has to be remembered that large volumes are often employed.

5.4.1 Distillation

The presence of organic impurities in a solvent will not necessarily directly influence an inorganic analysis. High levels of impurities can, however significantly affect the gross physical and physico-chemical properties of the solvent, indirectly affecting the analytical performance and batch to batch reproducibility. The use of pure organic solvents is therefore sometimes necessary more to ensure the uniformity of batches of reagents than to reduce the trace impurities in the solvent.

Conventional distillation by boiling is particularly suitable for the purification of solvents but it is frequently not 100% efficient. Under most circumstances the more sophisticated designs of distillation apparatus, with their associated reduced turnover, can be employed, as volume requirements are generally less than with water purification. The advantages and disadvantages of distillation for the purification of water are discussed elsewhere in the book. Aerosol formation resulting from the boiling action leads to droplets of unpurified material being carried over into the distillate, but this can be minimized by apparatus design and careful operation. Carry-over in this manner is reduced by smooth boiling and an efficient trap for droplets should be provided. The more tortuous the path between boiler and receiver, by using for example a column filled with silica helices, the greater should be the efficiency of distillation and the trapping of droplets. At the same time, this will unfortunately result in a reduced yield. The use of a high efficiency still will result in the removal of a high proportion of both the organic and inorganic impurities, and solvents of greater than 99% purity can often be obtained.

The sophistication of the fractionation apparatus to be employed varies between applications. In some, the main role of the distillation may only be to remove involatile inorganic contaminants resulting from the production and storage of the solvent. For this a simple fractionation column may be all that is required, with the initial and final 10% fractions of the distillate being discarded. In the case of the distillation of ethers, special precautions should be taken to remove peroxides before the distillation is commenced. It is recommended that the last 25% of the solvent should not be distilled to minimize the explosion risk arising from overheating the potentially explosive residue. It must be remembered that the quality of the distillate will depend largely on the speed of distillation and that inorganic impurities, while generally having boiling points well above those of the organic solvents, are most frequently carried over in the distillate in a fine mist of the impure solvent.

The most suitable apparatus for the everyday distillation of higher purity solvents is a fractionation column approximately 1 metre in length, having an internal diameter of approximately 2.5 cm and packed with borosilicate glass helices. However, such apparatus is not ideal if the solvent is to be employed in the analysis of any of the major elements found in the borosilicate glass. Under these circumstances silica stills may be appropriate.

For the preparation of some higher purity materials it is possible to employ subboiling distillation. This technique has been more widely applied to the purification of inorganic reagents, however, and is dealt with in greater detail later in the chapter.

5.4.1.1 Removal of inorganic impurities by pretreatment

Surprising quantities of inorganic impurities can be carried over during the distillation of a solvent. Although, for example, the lead content of chloroform can be significantly reduced by distillation, an additional order of magnitude improvement in the purity level can be obtained by first washing the solvent with high purity hydrochloric acid (ca. 2 mol/dm^3). Rinsing of excess dissolved acid from the solvent with water is advisable prior to distillation, especially if an acidic distillate is to be avoided.

5.4.1.2 Removal of organic impurities by pretreatment

If a distillation step is incapable of separating an impurity from the solvent, or, if it is preferable to remove the impurity before distillation, a pretreatment step can frequently be employed. The following are some examples of approaches which can be taken to reduce impurities prior to distillation.

Purification of alkane solvents such as hexane and cyclohexane

Some of the most commonly employed solvents are the alkanes and cycloalkanes. The effectiveness of their distillation can be significantly improved by nitration or sulphonation of impurities in the raw solvent. This is carefully carried out under reflux for some hours.

Nitration of aromatic or other susceptible impurities significantly raises their boiling points, improving the ease with which they can be removed by distillation. As the nitration proceeds, the nitrating mixture goes brown and is periodically replaced until, after reflux, it is no longer significantly coloured. The solvent is then rinsed with pure water and redistilled.

A sulphonation approach is similar, but the products of the reaction are mostly charged and can be washed from the solvent with water. Subsequent distillation removes residual impurities.

Whilst the mechanisms behind the effectiveness of both of these methods lie with an alteration of the chemical structures of the impurities, an additional benefit for trace metal analysis is that both procedures expose the solvent to highly acidic conditions, resulting in the partition of metals out of the solvent and into the acid phase. The rigour of the reflux step ensures that even the most insoluble inorganic impurities, which are often of a particulate nature, are brought into solution.

Purification of ethers

Linear and cyclic ethers such as diethyl ether, tetrahydrofuran and 1,4,-dioxane form peroxides when left in contact with the air. For this reason reagent grade chemicals are normally supplied containing a stabilizer. This is removed during purification, leaving the solvent susceptible to oxidation and the formation of peroxides. Distillation of ethers containing peroxides poses the risk of explosion, and peroxides must therefore be shown to be absent by a starch/iodide test[†] or removed before distillation. There are a number of ways to destroy the peroxides. Means which have been employed for the purification of diethyl ether include shaking with alkaline potassium permanganate for several hours, shaking with acidic ferrous sulphate solution, allowing to stand over copper powder or iron filings, shaking with a zinc/copper couple or solid ferrous sulphate, and passing the solvent through an alumina column. Most methods then involve washing the solvent with water and drying over calcium chloride. It must be emphasized that these purified solvents are susceptible to aerial oxidation and must be stored in the dark, preferably under nitrogen. They should not then be stored for more than a few weeks.

5.4.1.3 Drying the solvent

The separation of some impurities from a solvent by distillation may not be possible without additional treatment. The most commonly encountered impurity, and especially so after aqueous pretreatment methods have been employed for preliminary purification, is water. It may sometimes prove to be possible to carry out the distillation in the presence of a drying agent, but the impurities introduced in the

† To test for significant levels of peroxides, add 1ml of a 10% (w/v) solution of potassium iodide containing a drop of starch indicator to 10 ml of ether in a closed flask. If colour develops in one minute, peroxide is present and must be removed or destroyed before distillation.

process may themselves be unacceptable. Any drying of the solvent should normally be carried out prior to the distillation step as the addition of drying agents to the purified product will almost inevitably contaminate. Once it has been dried, the solvent should be stored in a sealed container to minimize the ingress of atmospheric water. Careful consideration of the requirements of the analysis may reveal that the complete removal of water from a solvent to be used in the trace analysis of an already wet sample is unnecessary.

If a drying agent is to be employed, its chemical nature must be considered to ensure that it is compatible with the material to be dried and that its use will not significantly influence the subsequent analytical procedure. Shown below (Table 5.2) are some possible drying agents for the small set of common solvents which are frequently employed in analytical procedures.

Table 5.2 Drying agents (Y = suitable) 1 = $CaSO_4$; 2 = $MgSO_4$; 3 = Na_2SO_4; 4 = CaO; 5 = K_2CO_3; 6 = $CaCl_2$; 7 = P_2O_5; 8 = Na

	Drying agent							
Compound	1	2	3	4	5	6	7	8
Organic acids	Y	Y	Y					
Alcohols	Y	Y		Y	Y			
Alkyl halides	Y	Y	Y			Y	Y	
Ethers	Y	Y				Y		
Hydrocarbons	Y	Y				Y	Y	Y[a]
Ketones	Y	Y	Y		Y			

[a] Saturated hydrocarbons only.

Solvents which form azeotropes with water present a particular difficulty, which can be overcome by distillation of the solvent in the presence of a desiccant.

5.4.2 Freezing Techniques

When a liquid freezes, impurities collect at the solid–liquid interface. If the freezing is carried out slowly, this can result in a simple but significant improvement in the purity of the frozen liquid. Dioxane, for example, can be purified by slowly freezing the liquid at 5 °C while stirring the unfrozen material. Once 90% of the dioxane has frozen the supernatant liquid containing the impurities can be siphoned off and discarded. Repeating the process results in further purification. The method is also suitable for acetic acid (f.p. 16.6 °C), aniline (f.p. −6°C), carbon tetrachloride (f.p. −23 °C), cyclohexane (f.p. 6.5 °C), cyclohexanone (f.p. −16 °C) and *p*-xylene (f.p. 14 °C).

5.4.3 Solvent Partition

In its simplest form a solvent extraction employs two immiscible solvents, one of which is normally aqueous. The equilibration of these two solvents by shaking results in the partition of dissolved material between the two phases. If the correct conditions are chosen this can be an effective means of purifying either phase.

Consider the equilibration of a solvent such as diethyl ether with a dilute acid. Under such conditions basic material is protonated to give positively charged species, which move into the acid phase. At the same time the protonation of acidic impurities gives uncharged products, which move into the organic phase. If the acid phase is then replaced by a dilute solution of a strong base, the acidic impurities are deprotonated (ionized) and move from the ether into the aqueous phase. The net result of these two steps is that both acidic and basic impurities have been removed from the ether. A side effect of the process will be the extraction of many inorganic impurities from the solvent.

This is a simple but effective way of removing some types of metal contamination from water-immiscible solvents. Whilst care must be taken to ensure that excessive pressure build-up does not occur, solvent purification can be carried out in a simple separating funnel. For the removal of metals from a solvent, acid extraction is generally all that is required. Typically, the solvent is equilibrated with approximately 3 mol/dm^3 acid in a solvent:acid ratio of 10:1. The acid is discarded and replaced by fresh material, re-equilibrated and then discarded. The solvent must then be washed with high purity water to strip out residual acid. Clearly this approach is not readily applicable to solvents which can dissolve substantial amounts of aqueous reagents, such as 4-methyl-2-pentanone, from which it can be very difficult to remove all acid residues. The acid should be carefully chosen to ensure that the important metal impurities are soluble in it.

5.5 PURIFICATION OF LIQUID INORGANIC REAGENTS AND SOLUTIONS

5.5.1 Distillation

5.5.1.1 Conventional distillation

With the exception of the purification of water, which is dealt with elsewhere, and a limited number of other compounds, conventional distillation is not widely applicable to the purification of inorganic reagents.

5.5.1.2 Isothermal (or isopiestic) distillation

A very simple procedure which can be used for materials having a high vapour pressure at room temperature (such as ammonia solution, acetic acid and hydrochloric acid) is isothermal distillation. For the purification of ammonia solution, equal volumes of raw ammonia solution (specific gravity 0.88) and high purity water are placed in separate containers held in a tightly closed container at room temperature (Figure 5.1). Over a period of several days the ammonia

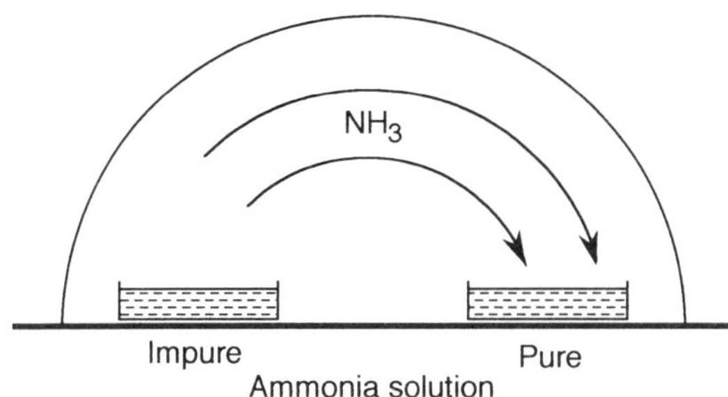

Figure 5.1 The purification of amonia solution by isothermal distillation

equilibrates between the two aqueous solutions. The ammonia transfers to the clean water, leaving behind involatile trace elements and generating a high purity ammonia solution having a concentration of approximately 7 mol/dm^3.

The purification of hydrofluoric acid by this means should be carried out in HF-resistant materials such as PTFE.

The concentration of the purified reagent after a single isothermal distillation should be approximately 50% of the concentration in the starting material. To increase the concentration of the purified material the original reagent should be replaced every two days. The concentration of the final product should approach the average of the concentration present in the raw and the purified material in the chamber.

5.5.1.3 Sub-boiling distillation

Probably the most effective means of obtaining concentrated high purity acids and solvents is the method of sub-boiling distillation (Kuehner *et al.*, 1972). The purification is based on the vaporization of the liquid by radiative heating of the surface of the liquid to prevent boiling. The efficiency of the method is derived from the distillation of material without the formation of spray or droplets. In the conventional distillation process the violent boiling action of the still leads to significant carry-over of raw solvent with the distillate. Surface evaporation in the sub-boiling still does not generate such a spray and the condensate is therefore of higher purity.

Provided that the reagent of interest is liquid and has a significant vapour pressure, the technique may be suitable. Examples of chemicals which have been shown to be amenable to this approach include—

acids: hydrochloric, nitric, acetic, hydrofluoric (PTFE still);

bases: ammonia solution;

solvents: chloroform, Freon TF, ethanol.

The effectiveness of the purification will depend on many factors, including the purity of the starting material and the cleanliness of the still system. An indication of the purity of reagents that can be obtained by this method is given in Table 5.3.

Table 5.3 Impurity levels in hydrochloric acid obtained from sub-boiling distillation

Metal	ng/g	Metal	ng/g
Fe	1.0	Ca	0.06
Na	1.0	Sn	0.02
Mg	0.2	Ba	0.04
K	0.5	Ag	0.005
Cr	0.04	Cd	0.005
Ni	0.3	Tl	0.005
Zn	0.02	In	0.001
Cu	0.05	Sr	0.005
Pb	0.01	Te	0.01

A particularly simple design of still described by Mattinson (1972) consists of two FEP bottles connected at right angles by a PTFE block. One bottle containing the reagent to be purified (Figure 5.2) is heated by an infrared lamp, and the condensate accumulates in the other bottle.

An advantage of this system is that it is totally enclosed and can be cleaned by simply refluxing acid in the inverted bottles. As it is made of fluorocarbons it is appropriate to the purification of hydrofluoric acid. Care must however be taken due to the pressure build-up in the bottles during use.

Most sub-boiling stills are constructed of fused quartz or silica, and a typical example is shown in Figure 5.3.

The unit is constructed entirely of high purity silica and heating is carried out by silica-sheathed elements above the liquid. The inclined cold finger condenses the vapour and leads the condensate into a collecting vessel outside the still. Normally a constant head device feeds the still reservoir and an overflow tube can be incorporated to prevent the stock reagent contaminating the distillate. Distillation rates are typically much slower than with conventional distillation, and the still reservoir has to be periodically flushed owing to the build-up of contaminants.

5.2.2 Filtration

Anyone who has looked closely at a concentrated solution of a reagent will have noticed that normally not all the material has dissolved. Provided that the solubility of the reagent has not been exceeded, this will be due to the presence of insoluble substances ranging from paper fibres to chemical impurities. In the case of charged complexants these impurities may well be insoluble metal complexes. As many purification procedures require the preparation of a solution of the reagent, it

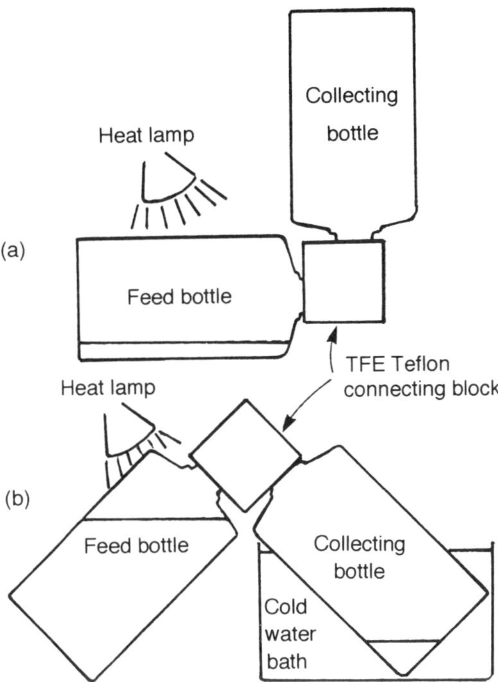

Figure 5.2 A PTFE sub-boiling still (Mattinson,1972). (a) Washing the apparatus under reflux. (b) Distillation. Reprinted with permission from Mattinson, *Anal. Chem.*, **44**,1715. Copyright (1972) American Chemical Society

is a very simple matter to filter the solution using a membrane filter with a nominal cut-off of 0.2 μm prior to more extensive purification.

5.5.3 Co-precipitation

The co-precipitation of trace impurities can be employed to remove many metals and non-metals from aqueous solution. The most commonly used procedure of this type involves the precipitation of iron, titanium or lanthanum carriers from solutions of reagents having a low tendency to form oxide or hydroxide precipitates themselves. Such an approach is particularly suitable for the purification of a number of alkali or alkaline earth salts.

The removal of iron from calcium nitrate solution, for example, can be achieved by the addition of lanthanum nitrate or aluminium chloride solution to an aqueous solution of the material to be purified. Oxidation of the iron to its iron(III) state with a few drops of hydrogen peroxide solution is then followed by a pH adjustment (to 7.5 for the lanthanum carrier or 6.0 in the case of aluminium) using

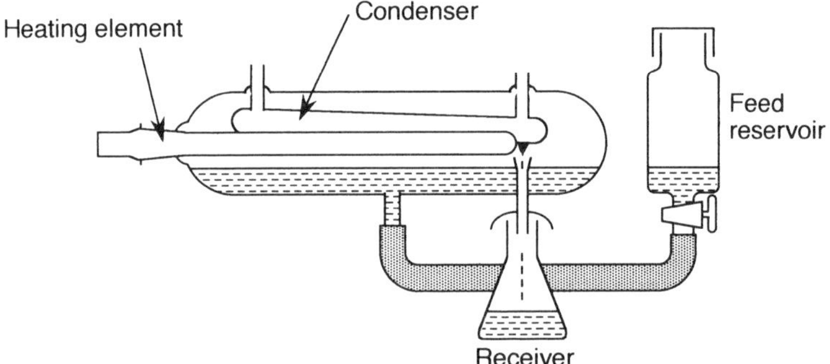

Figure 5.3 A quartz sub-boiling still (after Kuehner *et al.*, 1972)

ammonium hydroxide solution. Heating to 60 °C completes the co-precipitation of the carrier hydroxide with the insoluble iron(III) hydroxide. After cooling to room temperature the precipitate containing the iron can be removed by filtration.

Other examples of co-precipitation techniques which can be employed in reagent purification include:

(1) the removal of lead as its sulphate by co-precipitation with barium sulphate;

(2) the precipitation of cadmium sulphide with indium sulphide;

(3) removal of nickel by coprecipitation with copper dimethylglyoxime;

(4) the co-precipitation of phosphate with iron(III) hydroxide.

In devising a new co-precipitation method for the purification of reagent solutions, it is important to consider the chemical compatibility of the material to be purified in the context of the proposed procedure. The sulphide precipitation procedure precipitates a wide range of heavy metals from solution but would not be appropriate for the removal of cadmium from a zinc standard. In addition, the nature of impurities left in solution after the treatment must be considered. It is rarely possible to produce a product which is totally free of the chemicals which have been employed in the co-precipitation procedure.

5.5.4 Electrochemical Methods

If a controlled potential is applied across a solution using a mercury cathode and a platinum anode, many trace metal impurities are reduced at the mercury electrode (Figure 5.4). Electro-reducible impurities collect in and on the mercury cathode and can be subsequently removed.

A restricting aspect of the technique is that it is necessary for the solution to be

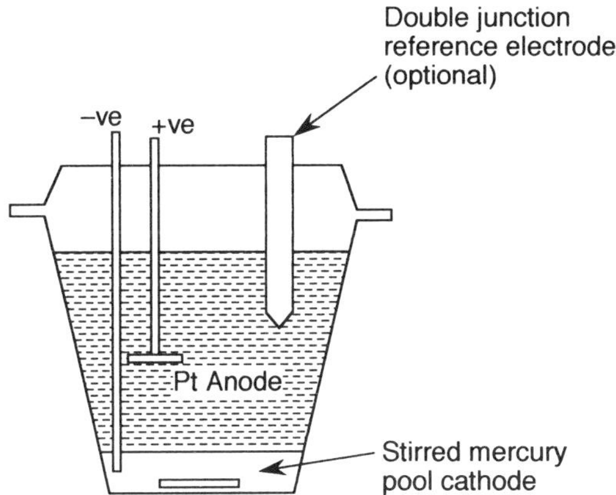

Figure 5.4 An electrolysis cell for the purification of salt solutions

conducting, and this method is therefore best suited to the purification of salt solutions which are not themselves reducible at the mercury electrode. Since pure water is a poor conductor of electricity the technique is unsuitable for the purification of water itself. In the right application, however, electrolysis can produce very pure aqueous solutions, and the technique has found many applications, particularly in the purification of buffers and supporting electrolytes for differential pulse polarography and anodic stripping voltammetry.

The potential is applied across the two electrodes and the electrolysis is allowed to proceed for some hours before removal of the electrodes. The working electrode potential can be better controlled by incorporating a reference electrode (silver/silver chloride or calomel) into the system, but care must be taken to ensure that electrode leakage does not itself contribute significantly to the trace metal concentrations which are being removed. A double junction reference electrode may help to overcome such problems.

The potential which must be applied to the cell depends on the reduction potential of the element and its concentration in solution. An estimate of the minimum potential which will be required to remove a particular element from solution can be obtained from the Nernst equation and tables of standard electrode potentials.

5.5.5 Ion-exchange

Ion-exchange can be exploited for the removal of charged impurities from reagents provided that sufficient selectivity is available. The most commonly encountered ion-exchangers are resin based and are therefore best suited to the purification of aqueous solutions and the less reactive reagents. They can be employed either for the collection of impurities or the removal of unretained impurities whilst the material of interest is held on the exchanger. Examples of applications include:

(1) the removal of iodine (as tri-iodide) from hydriodic acid by passing through a column of anion-exchange resin in its iodide form;

(2) the production of carbonate-free solutions of sodium or potassium hydroxides by passage over an anion-exchange resin such as Amberlite IRA-400 in its hydroxyl form.

More specific removal of metals can be achieved by the use of chelating resins such as Chelex-100 or Dowex A-1. These are particularly effective at removing heavy metals from alkali metal salt solutions (including sodium sulphide). The properties and preparation of ion-exchange resins are discussed in greater detail later in this chapter.

5.5.6 Solvent Extraction

Solvent extraction can be employed directly for the removal of soluble impurities or can be used in combination with complexing agents. Typical examples include:

(1) the removal of copper from potassium dihydrogenphosphate solution by extraction with diphenylthiocarbazone in carbon tetrachloride;

(2) the purification of an aqueous solution of *n*-decyl sulphate by continuous extraction with diethyl ether;

(3) the Soxhlet extraction of sodium dodecyl sulphate with petroleum ether;

(4) the extraction of iron from sodium thiocyanate solutions with diethyl ether.

Problems which can be encountered in the use of such extraction procedures are largely derived from the presence of residual extractants and complexing agents in the products.

5.6 PURIFICATION OF SOLID REAGENTS

The techniques which are described in this section are those which are designed to produce a solid product. They are in the main techniques which have their origins in chemical synthesis and manufacture and are therefore directed towards the purification of material to a high percentage purity; they do not necessarily produce materials which have particularly low levels of a particular constituent. They are particularly useful in the preparation of pure materials for the production of analytical standards where it is the percentage purity which is the important criterion. They can, however, sometimes be suitable for reducing the levels of trace impurities.

5.6.1 Sublimation

The process of sublimation is analogous to distillation, but in this case the transition involved is from the solid to the gas phase. Confusingly, both the vaporization of a solid and the combined process of solid vaporization and condensation of the vapour (cf. distillation) are called sublimation. If a solid phase element or compound has a significant vapour pressure it is possibly suitable for purification by this process.

The apparatus which is required is simple to construct, consisting of a chamber which holds the impure solid material and a condenser to collect the purified solid (Figure 5.5).

The rate at which the material is sublimed can be increased by:

(1) heating the impure material while ensuring that it does not melt or decompose;

(2) reducing the partial pressure of the vapour in the system by applying a vacuum or sweeping the vapour away from the impure material with a carrier gas;

(3) reducing the distance between the evaporator and the condenser;

(4) increasing the surface area of the sublimand.

Amongst those substances which are amenable to purification by sublimation-are: aluminium, chromium, ferric, titanium, uranium and zirconium chlorides; calcium and magnesium; iodine and sulphur; salicylic acid and pyrogallol. The

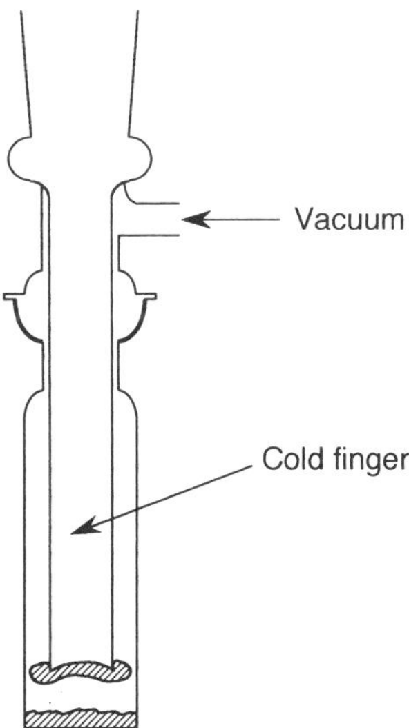

Figure 5.5 Equipment for the purification of reagents by vacuum sublimation

purification of 8-hydroxyquinoline, for example, can be performed by the vacuum sublimation at 70 °C of material which has been recrystallized from ethanol (95%).

Preparation of a pure reagent may not involve the purification of the reagent itself

It is worth noting that the most effective means of producing a pure reagent may not be to start from the reagent itself but to prepare it from one which is more readily purified. An effective method for the preparation of phosphoric acid, for example, involves the purification of phosphorus pentoxide by sublimation followed by the dissolution of the oxide in pure water.

5.6.2 Precipitation and Recrystallization

Precipitation and recrystallization are alternative approaches to the deposition of material from solution. In each the dissolved material is deposited in a form which may be purer than the original reagent, as some of the impurities remain in solution. In both cases the deposition may be induced by the addition of an extra reagent. Precipitation is the faster of the two but, as the material is often deposited in an amorphous state, the purity of the product may not be as high as that achieved with a material which has adopted a crystal structure.

A common example of the application of a precipitation procedure is the purification of the versatile chelating agent ethylenediaminetetraacetic acid (EDTA). This reagent is commonly available as its soluble disodium salt but it can be used in most applications as the acid. Purification of the EDTA is carried out in two stages, the first being the removal of insoluble impurities by filtration (0.2 μm membrane) of an aqueous solution of the reagent. The resulting solution is then slowly acidified with high purity mineral acid. This results in the displacement of metals from the complexing agent and precipitation of the pure free acid.

Recrystallization is one of the most commonly employed methods for the purification of solid materials. The method relies on the exclusion of contaminants from the crystal lattice and is best suited to the removal of those impurities which cannot be incorporated in the lattice owing to, for example, size or charge incompatibilities. Crystallization occurs from a supersaturated solution of the material to be purified. As solubility increases with temperature, a saturated solution is prepared at elevated temperature and is allowed to cool. Crystal growth will then start on any suitable nucleating surface. This is often a speck of dust, but in trace analysis this should not be present!

In order to maximize the exclusion of impurities, the recrystallization process should be performed as slowly as is practical. Recrystallization should therefore be carried out from a solvent in which the material to be purified has only moderate solubility. If the compound is sufficiently stable, gentle heating may be employed to increase solubility to allow a supersaturated solution to be prepared. For a slow recrystallization to occur, the extent of supersaturation should be limited and the solution should be allowed to cool to room temperature naturally. Seeding of the solution with pure crystals of the material will assist in the recrystallization process. Unless absolutely necessary, the solution should not be refrigerated or moved while recrystallization takes place. In the event of crystals having failed to form within a few days, scratching the container walls with a glass rod or gentle movement of the container may initiate crystal growth. If this fails refrigeration may help. The resulting crystals should be rinsed carefully after they have been formed to remove impurities which collect on the surface and concentrate in the solvent during crystallization. It should be noted that the solvent can become incorporated into the crystal structure or the number of waters of crystallization may change. The resulting crystals may thus not have the same chemical composition as the starting material.

Those compounds which are thermally unstable can be dissolved at room temperature and recrystallized in a refrigerator.

5.6.3 Zone Melting

If an impurity is less soluble in a crystalline solid than in a melt of that material, significant improvements in purity can be obtained by zone refining (or zone melting). A tube (Figure 5.6) is filled with the impure material and a narrow band of

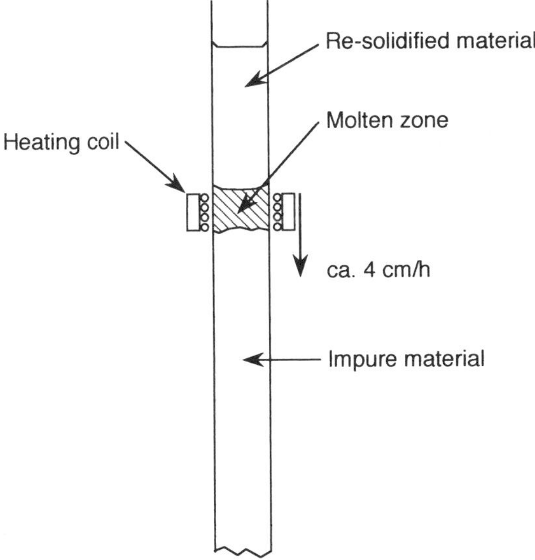

Figure 5.6 Zone melting apparatus

material is melted by a heater which slowly advances along the tube. At the front of the heated region the impure material is melting and at its back pure material is solidifying. As the molten band progresses along the tube the concentration of impurities therefore increases in the molten region. This leads to an increasing impurity level within the purified region and purity is therefore not constant throughout the tube. Several passes of the heater are therefore normally carried out to enhance the purity of the product.

5.7 ION-EXCHANGE RESINS

Ion-exchange resins play an important role in many trace analyses. Not only are they employed in the purification of reagents such as water; they are particularly useful for the preconcentration of the analyte and its separation from interfering species.

The most commonly encountered ion-exchangers consist of irregular, three dimensional networks of hydrocarbon chains. With the exchangers (Figure 5.7) which are based on the copolymer of styrene with divinylbenzene (DVB), the proportion of DVB governs the degree of cross-linking of the polymer and the extent to which the resin can swell during use.

High degrees of cross-linking (e.g. 20% DVB) make the resin less susceptible to swelling but more brittle. The diffusion of ions through the resin is retarded by the highly cross-linked polymer matrix and hence the exchange is faster for resins with less cross-linking. Resins with little cross-linking are not physically strong yet offer the highest rates of exchange. To circumvent this conflict, macroporous resin beads have been developed (Figure 5.8) with each ion-exchange particle consisting of bonded microspheres of highly cross-linked polymer. This results in the presence of large pores through which ions can readily move with a polymer matrix having the mechanical strength of a highly cross-linked resin.

5.7.1 Types of Resin

There is a wide selection of solid ion-exchangers available but all consist of a solid core with a chemically reactive surface. The ion-exchangers fall into three main categories: cation, anion and chelating exchangers.

5.7.1.1 Cation-exchange resins

The surface of a cation-exchange resin is typically modified with sulphonic acid groups (Figure 5.9).

These groups are charged and must be associated with a counter ion to ensure the overall charge neutrality of the resin bead. A resin having sodium as counter ion is described as being in the 'sodium form'. It is the counter ion which is the

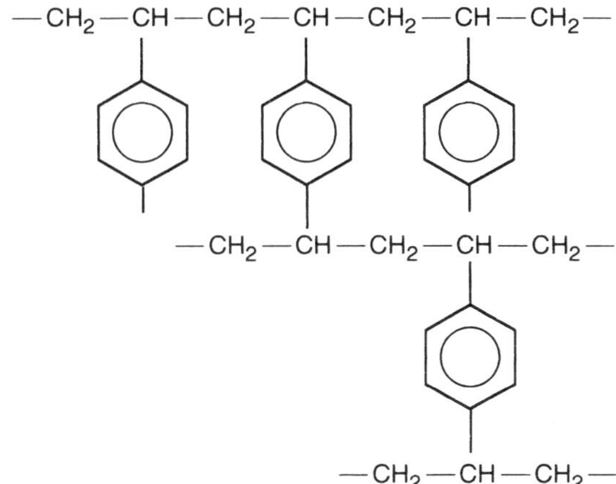

Figure 5.7 The styrene–divinylbenzene copolymer structure

basis of chromatography as cation or anion collection by the resin results from the exchange of the counter ion for another ion.

$$\text{RESIN—(H)} + \text{Na}^+ \rightleftharpoons \text{RESIN—(Na)} + \text{H}^+$$

The cation-exchange resins are further divided into two groups, the strong and weak acid exchangers, which differ according to the nature of acidic group on the resin surface (Table 5.4).

5.7.1.2 Anion-exchange resins

An anion-exchange resin has on its surface positively charged moieties, such as the quaternary ammonium group (Figure 5.10). There are two main types of strong anion-exchange resins, categorized as Type 1, in which the functional group is

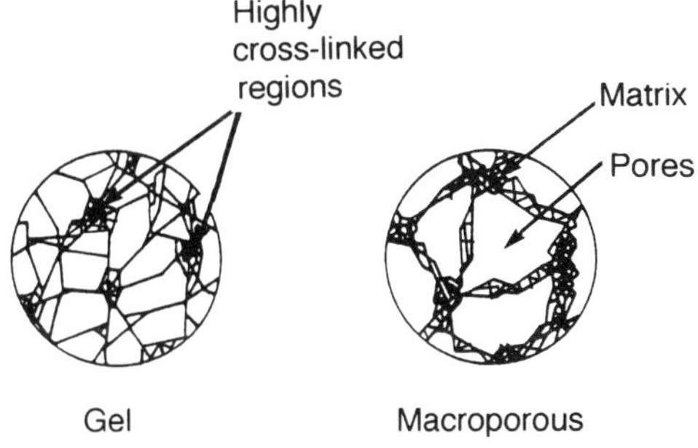

Figure 5.8 Microporous gel and macroporous resin beads

Table 5.4 Some typical cation-exchange resins

	Dowex[a]	Amberlite[b]	Diaion[c]	Lewatit[d]	Permutit[e]	Bio-Rad[f]
Strongly acidic polystyrene $-SO_3^-H^+$	50W HCRW	IR-116,118 120,77,218 163,169,122 124,130,140 169	SK102,103 104,106,1A,1B 110,112,116	PN S100	Zeocarb 225	AG 50W
Strongly acidic macroporous $-SO_3^-H^+$	MSC-1	200, 252 Amberlyst 1S	PK204,208 212,216,220 224,228			AG MP-50
Weakly acidic $R-COO^-Na^+$	CCR-2	IRC-84,50,72 CG-50,DP-1 IRP-64	WK10,11	CN0	Zeocarb 226	Bio-Rex 70
Weakly acidic chelating $-CH_2N(CH_2COO^-)_2H^+{}_2$	A-1		CR10			Chelex 100 Chelex 20

[a] Dow Chemical Company;
[b] Rohm & Haas Co.
[c] Mitsubishi Chemical Co.
[d] Bayer
[e] Permutit Company
[f] Bio-Rad.

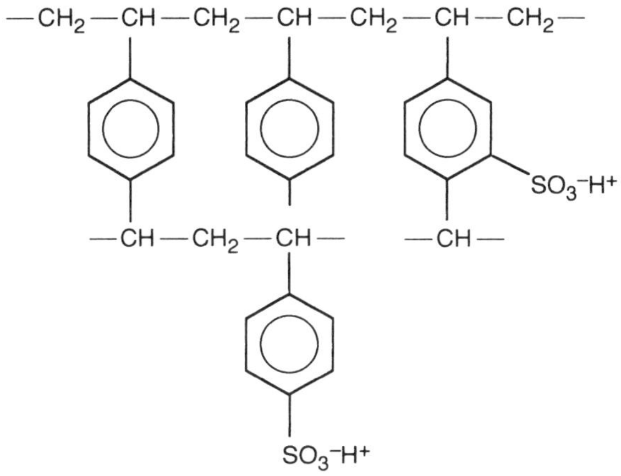

Figure 5.9 The structure of a monofunctional strongly acidic cation-exchange resin in the 'hydrogen form'

derived from a trialkylamine, and the Type 2 resins, formed from dimethylethanolamine. The Type 2 resins have lower capacities and lower basicity than the Type 1 resins and are chemically less stable.

In order to ensure the overall charge neutrality of the particle, the counter ion is in this case negatively charged.

$$\text{RESIN—(OH)} + NO_3^- \rightleftharpoons \text{RESIN—}(NO_3) + OH^-$$

The weak base resins have tertiary (-NR_2), secondary (-NHR) or primary (NH_2) amino groups. As they are weak bases the positively charged groups are formed by protonation, and they must therefore be used below pH 7.

> If it looks like plastic beads and smells of fish it is probably an anion-exchanger

None of the anion-exchangers is completely stable which accounts for the amine smell given off by the resin.

5.7.1.3 Chelating solids

The chelating solids are a group of materials having chelating or complexing groups on their surface. The most commonly encountered example of such a material is the chelating ion-exchange resin Chelex 100, which is based on the chemical

Table 5.5 Some typical anion-exchange resins

	Dowex[a]	Amberlite[b]	Diaion[c]	Lewatit[d]	Permutit[e]	Bio-Rad[f]
Type 1 Strongly basic polystyrene $-CH_2N^+(CH_3)_3Cl^-$	1 11	IRA 400,401,402 416, 420,430	SA 10A,10B 11A,11B,100	M500,5020 5080	Zerolit FF Permutit S-1	AG1
Type 2 Strongly basic polystyrene $-CH_2N^+(CH_3)_2(C_2H_4OH)Cl^-$	2	IRA 410	SA 20A,20B 21A,21B	M600	S-2	AG2
Type 1 Strongly basic macroporous $-CH_2N^+(CH_3)_3Cl^-$	MSA-1	IRA 900,904,938 958 Amberlyst A-26, A-27	PA 304,306 308,310,312 318,320	MP5080		AG MP-1
Type 2 Strongly basic macroporous $-CH_2N^+(CH_3)_2(C_2H_4OH)Cl^-$	MSA-2	IRA 910 420	PA 404,406 408,410,412 414,416,418			
Weakly basic polystyrene, phenolic or polyamine $-CH_2N^+(R)_3Cl^-$	WGR WGR-2	IRA 45,48,47 68,60,58	WA10,11	MIH	G	AG3 AG4
Intermediate/weak base macroporous $-CH_2N^+(R)_3Cl^-$	MWA-1	IRA 35,93 94,99 Amberlyst A-21	WA 20,21,30	MP7080		

[a] Dow Chemical Company
[b] Rohm & Haas Co.
[c] Mitsubishi Chemical Co.
[d] Bayer
[e] Permutit Company
[f] Bio-Rad.

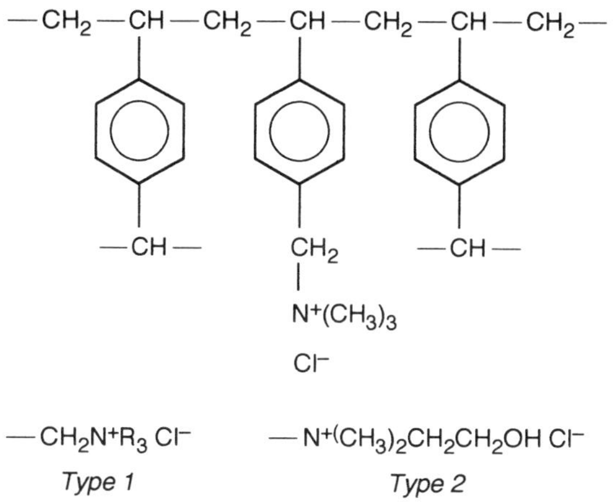

—CH_2—CH—CH_2—CH—CH_2—CH—CH_2—

—CH— CH_2 —CH—

$N^+(CH_3)_3$

Cl^-

—$CH_2N^+R_3$ Cl^- —$N^+(CH_3)_2CH_2CH_2OH$ Cl^-

Type 1 *Type 2*

Figure 5.10 The structure of a monofunctional Type 1 strongly basic anion-exchange resin in the 'chloride form' and the differences between Type 1 and Type 2 resins

modification of a resin surface with iminodiacetic acid groups (Figure 5.11).

This resin therefore has metal chelating properties which are similar to those of EDTA. In recent years a wide selection of these materials has been developed, based not only on resin cores but also on more rigid materials such as controlled pore glass and silica. The materials have two main advantages over conventional

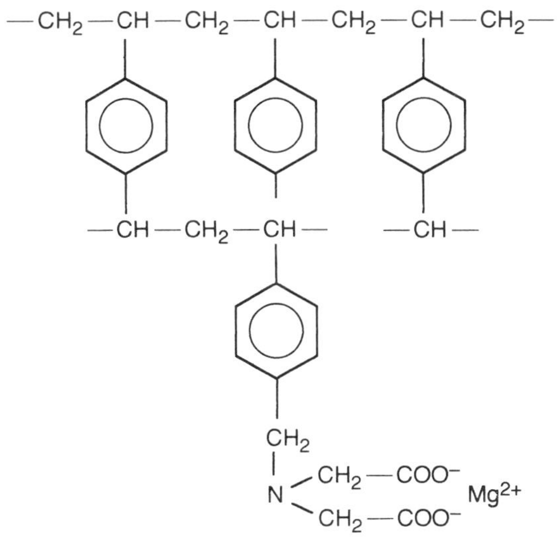

—CH_2—CH—CH_2—CH—CH_2—CH—CH_2—

—CH—CH_2—CH— —CH—

CH_2

N (CH_2—COO^-)(CH_2—COO^-) Mg^{2+}

Figure 5.11 The structure of a chelating resin in the 'magnesium form' chelating through the iminodiacetic acid groups

ion-exchange materials. They benefit from the chelate effect resulting in strong bonding of some metals and they can be partially selective in the metals which are bound.

Chelating solids differ from the typical anion- and cation-exchangers in that they are capable of binding metals through multiple coordinating groups attached to the surface. Whereas in some cases the metal binding action may involve an ion-exchange process, the chelation process may involve binding through lone pairs on coordinating groups such as polyamines. The multiple coordination of metals on a chelating resin makes it particularly suitable for the collection of polyvalent metal ions from solutions containing monovalent ions. By careful selection of the complexing groups on the resin, selectivity can be enhanced towards an element of interest, and the resin can be used for the chromatographic separation of metal mixtures.

5.7.2 Properties of Resins

The particle size of a resin determines its efficiency and the time required for it to reach equilibrium. The smaller the average particle size, the more rapidly will the resin come to equilibrium with species in solution and the more efficient will be its separation ability. For use in column chromatography, 100–200 mesh particles are generally recommended. The larger particle size material is generally reserved for batch applications and roles in which a critical separation of species is not required.

The capacity of a typical cation- or anion-exchange resin is approximately 5 millimoles of a monovalent ion per gram of dry resin. As polyvalent ions occupy more than one charged site on the resin surface, the capacities are correspondingly reduced to approximately 2.5 mmol/g for a divalent ion and 1.7 mmol/g for a trivalent ion. These capacities are often quoted as milliequivalents per gram.

The pH range over which an ion-exchange resin can be used is governed by the acid or base strength of the surface groups. Ion-exchangers having strongly acidic or basic groups are ionized over a wide pH range and are therefore useful over an equally wide range of acidity conditions. The weak acid or base exchangers, on the other hand, will be ionized over a more limited pH range, and their useful working range will be governed by the pK_a of the functional group (Table 5.6).

Table 5.6 Typical operating ranges of ion-exchange resins

	pH	
	Lower	Upper
Strong acid	1	14
Weak acid	5	14
Strong base	1	14
Weak base	1	7

5.7.3 Selectivity

Ion-exchange resins are selective and not all ions are held equally by a resin. For ions of the same charge, affinities are normally related to the radii of the hydrated ions. Polyvalent ions are attached more firmly than monovalent ions.

For strongly acidic cation-exchange resins the following affinity series generally hold (Table 5.7).

Table 5.7 The relative selectivities of the cation-exchange resin AG 50W-X8 (Bio-Rad Laboratories, 1982)

Counter ion	Relative selectivity	Counter ion	Relative selectivity
H^+	1.0	Fe^{2+}	2.55
Li^+	0.85	Zn^{2+}	2.7
Na^+	1.5	Co^{2+}	2.8
NH_4^+	1.95	Cu^{2+}	2.9
K^+	2.5	Cd^{2+}	2.95
Rb^+	2.6	Ni^{2+}	3.0
Cs^+	2.7	Ca^{2+}	3.9
Cu^+	5.3	Sr^{2+}	4.95
Ag^+	7.6	Hg^{2+}	7.2
Mn^{2+}	2.35	Pb^{2+}	7.5
Mg^{2+}	2.5	Ba^{2+}	8.7

The following series frequently holds for strongly basic anion-exchangers (Table 5.8)

Table 5.8 The relative selectivities of the anion-exchange resin AG 1-X8 (Bio-Rad Laboratories, 1982)

Counter ion	Relative selectivity	Counter ion	Relative selectivity
OH^-	1.0	CN^-	28
F^-	1.6	Br^-	50
propionate	2.6	NO_3^-	65
acetate	3.2	ClO_3^-	74
formate	4.6	HSO_4^-	85
IO_3^-	5.5	phenate	110
HCO_3^-	6.0	I^-	175
Cl^-	22	citrate	220
NO_2^-	24	salicylate	450
BrO_3^-	27	benzene sulphonate	500
HSO_3^-	27		

The general 'selectivity rules' (Korkisch, 1989) for most ion-exchange resins are:

(1) selectivity increases with the degree of cross-linking in the resin;

(2) small hydrated ions are preferentially adsorbed;

(3) if the ionogenic group on the resin forms an ion-pair with the reacting ion the selectivity of the resin to this ion will be increased;

(4) substances which are poorly ionized in solution will be less well retained than compounds which are readily ionized;

(5) selectivity decreases with temperature.

5.7.4 Preparation of Ion-exchange Resins for Use

5.7.4.1 Purification of a styrene–divinylbenzene cation-exchange resin

There are two main aspects to the purification of an ion-exchange resin. Firstly, the resins are often made by manufacturing processes which leave them containing high levels of organic and inorganic impurities. These must be washed from the resin to prevent interference in the analysis. Secondly, the resin must be converted into the form required for the analysis.

The exact nature of the purification which is required depends largely on the nature of the application. The impurities which are derived from the manufacturing process may be difficult to remove fully within a reasonable time because of their slow diffusion through the cross-linked matrix. For most purposes the cation-exchange resin must be purged of both organic and inorganic impurities while being converted into a suitable form for the analysis. The resin is most commonly converted into either its 'sodium' or 'hydrogen' form. Most resins produced for general purpose application are supplied containing levels of metal impurities which make them unsuitable for use in trace metal analyses. Resins which have been employed in an analytical procedure are often no less pure than the new resin and the recycling of used resin is often possible.

The purification and regeneration of a resin are most effectively carried out in a column. For small quantities of resin this might be a tube 60 cm long and approximately 2.5 cm in diameter. One end of the tube is fitted with a coarse frit and stopcock. The resin is slurried with an excess of dilute acid (typically 1 mol/dm^3 hydrochloric acid). This displaces metal impurities from the resin and converts the resin to the 'hydrogen form'. The slurry is carefully poured into the column while allowing the liquid to drain from the bottom. The volume occupied by the resin in the column is called the bed volume. Care should be taken to ensure that at no stage does the resin bed dry out[†].

> Never allow a chromatography column to dry out during use

† If this occurs the column should be almost filled with water, capped and inverted to totally resuspend the resin.

The liquid can then be carefully drained from the column until the top of the resin bed is just covered with liquid. In order to ensure complete conversion of the resin, sufficient of the counter ion has to be passed through the resin bed. The quantity of counter ion solution which has to be used depends on the affinity of the ion which is to be displaced for the resin. As a general rule use one bed volume of a 1 mol/dm^3 counter ion solution for each unit difference in relative selectivity, with no less than two to three bed volumes being used. Residual counter ion solution is then removed by passing high purity water through the resin until no free counter ion is detected in the eluate.

More stringent procedures may have to be employed for special applications. Many metals have particularly high stability constants with the chelating agent ethylenediamine tetraacetic acid (EDTA), and the resulting metal complexes are generally negatively charged water soluble species. This agent can therefore be effectively employed in the removal of trace metals from the cation-exchange resin. The formation of metal–EDTA complexes is favoured by moderate pH. The removal of metals from the resin by substituting the acid in the previously described procedure with a solution of disodium EDTA, made mildly alkaline with sodium hydroxide, would be expected to strip most metals from the resin and result in the conversion of the resin into its 'sodium form'. The excess EDTA would then be removed by washing with pure water. If the resin were required in its 'hydrogen form', acid would have to be passed through the resin in between the EDTA treatment and the wash step.

An additional trick which can be useful in removing metal impurities from cation-exchange resins is to convert the cationic form of a metal impurity to an anionic form. This is best exemplified by the removal of chromium from a cation-exchange resin. To be held by a cation-exchange resin it is probable that the metal is present in its cationic form, Cr^{3+}. Elution of the resin with an alkaline solution of hydrogen peroxide would oxidize the metal to CrO_4^{2-}, which, being an anion, would be released from the resin into solution. Sulphonic acid cation-exchange resins would not be significantly attacked by the oxidizing agent.

5.7.4.2 Purification of a styrene–divinylbenzene anion-exchange resin

The pretreatment of anion-exchange resins follows closely that which is employed for a cation-exchange resin. The first step of the process is to remove, as far as possible, all residues from the manufacturing process or previous applications of the resin. The resin is normally washed with water and then treated with a high concentration solution of an anion which is to be attached to the resin. This might be, for example, sodium hydroxide solution (1 mol/dm^3) to convert the resin to the 'hydroxide form' or sodium chloride solution to convert it to its 'chloride form'. The quantity of counter ion solution which is required can be judged using the criterion used previously for the cation-exchange resin: one bed volume for every unit difference in relative selectivity, provided that it is not less than two to three bed volumes.

Table 5.9 Ion-exchange purification schemes

A		B		C	
Wash treatment	Form	Wash treatment	Form	Wash treatment	Form
Water	As supplied	Water	As supplied	Water	As supplied
HCl	Cl^-	HCl	Cl^-	HAc	Ac^-
NaAc	Ac^-	HAc	Ac^-	Water	Ac^-
Water	Ac^-	Water	Ac^-		

(Ac- = acetate, HAc = acetic acid)

The exact nature of the resin treatment will of course depend on the nature and extent of the purity required for a particular application. If metals are to be removed from the resin, acid washing may have to be carried out at an early stage. The sequence of treatments can significantly alter the purity and final 'form' of the resin. Consider the preparation of an ion-exchange resin which is required in its 'acetate form' and which should be purged of trace metals. There are at least three ways by which this could be achieved.

The choice of procedure will depend on the requirements of the analysis (Table 5.9). Process A uses a strong acid to displace anions from the resin and to dissolve metal impurities; at the same time the resin is converted to the chloride form. The following wash with sodium acetate converts the resin to the acetate form, but does so under moderate pH conditions. Here lies a potential problem: the conditions may be ideal for the collection of impurities from the sodium acetate onto the resin. Process B partially circumvents this problem by using acetic acid to convert the resin to the acetate form. Process C is the most rapid, employing the acetic acid to carry out all the resin cleaning and conversion to the acetate form. The simplicity of this dual role is to be commended but only if the acetic acid is found to be sufficiently strong an acid to clean the resin effectively.

There is of course potential for contamination of the resin during the pretreatment process. The resin is at its most vulnerable once it has been converted to its final form, and it is therefore particularly important that the water which is used for the final wash does not contain significant levels of impurity, as these will become concentrated onto the resin.

5.7.4.3 Purification of Chelating Solids

Although not all of these materials are strictly ion-exchangers, and many are not resin based, the processes which are employed in their purification are similar to those described above for the cation-exchange resins. It must be noted that these materials are generally chosen for their strong metal binding and the processes involved in stripping contaminating metals from the surface may have to be more rigorous than can be employed for the less specific resins. The strength of acid which is required to displace a metal from the surface may therefore have to be stronger. An alternative approach is to consider the chemical process which is involved in the binding of contaminating metals onto the surface and to seek to exploit that chemistry in the clean-up procedure. In the case of the purification of Chelex 100, for example, washing with the chemically similar EDTA may be effective.

5.8 GASES

Gases have a number of uses in trace analysis. In electrochemical systems, for example, the inert or stable gases such as nitrogen and argon are used for the removal of dissolved oxygen. Gases are used as carriers in gas chromatography and offer purity advantages which make them particularly suitable for the preparation of high purity reagents. The micro-electronics industry in particular has placed stringent demands on the purity of commercially available gases. Many grades are now available and very high purity materials can be obtained, but at a cost.

In an electrochemical analysis in which nitrogen is used to remove oxygen from solution, the concentrations of oxygen and volatile organic compounds in the gas might have to be kept well below 1 ppm, but water might not be an important consideration. Obtaining commercial nitrogen with this purity may not be as difficult as it is to maintain the purity of the gas as it is delivered to the experiment. Contamination problems which arise from the use of gases in trace analysis originate not only from the commercial material but also from handling difficulties.

Typical of the classes of contamination which are encountered when using gases are:

- gaseous impurities from incomplete purification;
- water, oil and rust from the manufacture;
- contamination from the transfer of the gas to the experiment (handling).

The most common means of storage of large volumes of gas is in compressed or liquefied form. The pressures involved necessitate that most gases be stored in metal cylinders, which are unlikely to be made of corrosion resistant materials unless a large premium is placed on the reagent costs. For most applications gas is dispensed from the cylinder through a regulator, the design of which may not be well suited to the requirements of trace analysis or which may have undergone corrosion in previous applications. There are a number of things which can be done to minimize contamination from gases used in an experiment. As with all such situations, it is necessary to make a number of decisions regarding the requirements of the experiment.

- Identify those contaminants which may be present in the gas and which would be expected to influence the analysis.
- Choose the grade of gas and storage container which best fits the requirements of the analysis, bearing in mind that in-line purification of the gas may be the best possible option.
- Select transit lines and fittings to deliver the gas to the experiment which minimize potential contamination.

In an ideal world ultra-high purity materials would be available for any particular application, but it may be necessary to accept that by the time the gas has been stored in a cylinder and dispensed some contamination is inevitable. In many experiments it will therefore be necessary to develop in-line systems which purify the gas immediately before use.

5.8.1 Gas Fittings and Connections

For most experiments the gas delivery system is a straightforward line running from the cylinder through a regulator to tubing fitted with taps. Potential contamination starts with the regulator. Regulators come in a number of designs, many of which employ polymeric pressure reduction diaphragms, which can introduce trace quantities of volatile organics and may be permeable to oxygen. Metal diaphragm regulators are available for more stringent applications. Careful selection of the materials used to construct the regulator is particularly necessary for the corrosive gases such as hydrogen chloride and fluorine. Conventional regulators are subject to severe corrosion from these gases unless chemically resistant metals such as monel are used in their construction.

Tubing used in the apparatus should be carefully selected for the particular application and should be thoroughly cleaned before use. For many applications, and in particular the installation of permanent gas lines, copper or stainless steel tubing is a good compromise between cost and performance. Before installation it is imperative that this is thoroughly degreased, as recent batches of domestic grade 4 mm i.d. copper tubing have been found to contain approximately 10 mg of dichloromethane-extractable material per metre of tubing. Chlorinated solvents, although effective degreasing agents, can attack metal tubing and are particularly inappropriate for cleaning copper tubing.

The following guidelines may be helpful in the construction of new gas lines.

- Choose the correct regulator for the application.
- Choose suitable tubing. Contamination with volatile organics and gas diffusion both into and out of the tubing must be borne in mind if plastic tubing is to be considered. Plastic tubing should not be used for permanent installations of lines containing flammable gases as they may melt and leak in the event of a fire.
- Clean the tubing thoroughly before installation. For metal tubing thoroughly wash with a degreasing solvent and then evaporate residual solvent out of the tubing with a stream of clean air or nitrogen.
- Use high quality fittings. For the connection of metal tubing these should have metal ferrules and should not require the use of Teflon tape.
- Taps and flow regulators frequently use polymeric components and contain grease, and these components should be carefully selected with the analytical application in mind.

- Include a particle filter in the line close to the point where the gas is used in the experiment. In-line particle filters before pressure and flow regulators can significantly extend their lifetimes.

5.8.2 Purification of Gases

5.8.2.1 Oxygen removal

A number of methods are available for the removal of oxygen from inert or stable gases. These fall into two categories: those employing aqueous reducing agents used in bubbler systems and methods which involve the oxidation of solids. Several laboratory methods employ scrubbing solutions; these can be based on the quantitative reactions of oxygen with alkaline pyrogallol (1,2,3-trihydroxybenzene) solution, sodium dithionite, chromium(II) or vanadous salts[†]. Oxygen can be removed from many gas streams by passing it through a tube filled with 'low sulphur' copper wire heated to 450–500 °C. With gases such as nitrogen, helium and argon, oxygen is removed by oxidation of the copper. With hydrogen, water is formed from the oxygen. Alternative solid phase systems for the removal of oxygen include the use of finely divided palladium at room temperature for the purification of hydrogen and manganese(II) oxide coated on alumina.

5.8.2.2 Drying gases

There are a number of chemical reagents which can be employed for the removal of water from gases. Some are more effective than others (Table 5.10) but the most efficient drying agent may not be the most appropriate in all circumstances because of potential contamination of the resulting gas stream.

Table 5.10 The effectiveness of drying agents: residual water content of air (Falbe and Hahn, 1974)

Drying agent	Water content (g/m^3) after drying at 25 °C
P_2O_5	0.00002
$Mg(ClO_4)_2$	0.0005
H_2SO_4 (100%)	0.003
Alumina	0.003
Molecular sieve 4A	0.01–0.0001
CaO	0.2
$CaCl_2$	0.14–0.25
H_2SO_4 (95%)	0.3
Silica gel	0.5

† To prepare a pyrogallol solution for scrubbing oxygen from a gas stream, dissolve 15 g of pyrogallol and 30 g of potassium hydroxide in 1 litre of water. The solution is effective above 18 °C

Vanadous solutions are prepared by boiling ammonium metavanadate (2 g) with 25 ml of concentrated hydrochloric acid, diluting to 200 ml and then shaking with a few grams of amalgamated zinc. This would obviously not be ideal for many trace metal analyses.

The most important criterion to employ in the selection of a suitable drying agent is the potential reactivity of the gas with the drying agent. Taking two extreme examples, the drying of an acidic gas by a basic drying agent such as calcium oxide would not be advisable, and readily oxidized materials should not be dried with sulphuric acid. The physical nature of the drying agent may also be of importance, as solids can be more readily incorporated into a pressurized gas stream than liquids. In all cases it must be remembered that, although water will have been largely removed from the gas, the drying agent may well have left a particulate or vapour residue in the 'purified' gas, necessitating yet further purification steps.

5.8.2.3 Other gases and vapours

It sometimes seems that as many methods have been devised for the purification of gases as there are combinations of gases and impurities. When considering the application of a particular method, it is important to consider both the chemical compatibility of the method with the gas to be purified and whether, in purifying the gas stream of one contaminant, others may be generated. Some methods remove a contaminant whereas others merely convert it into something else.

As with the removal of oxygen, solid phase adsorbents such as carbon, silica, alumina or the molecular sieves are frequently the most convenient for removing other impurities, their advantages being that they can be semi-permanently incorporated into a gas line, can be regenerated and do not generally introduce reaction products into the gas stream. With all solid phase purification systems, however, care must be taken to ensure that particulate matter which breaks off the adsorbent during use is filtered out of the gas stream before it is used. One of the most popular general systems for the purification of the more inert gases is the use of molecular sieves in combination with activated carbon. The molecular sieves provide a size-based removal of contaminants (Table 5.11) and residual material is adsorbed onto the activated carbon.

Table 5.11 Molecular sieves and gas adsorption

Type	Pore diameter	Molecules adsorbed
4A	0.4 nm	H_2O and CO_2, H_2S, SO_2, NH_3, C_2H_6 Below −30 °C: N_2, O_2 and CH_4
5A	0.5 nm	As 4A plus butane, hexane, n-butanol etc., but not branched chain C_6 hydrocarbons, cyclic hydrocarbons, secondary or tertiary alcohols, or CCl_4.General purpose drying/purification
13X	1 nm	Above plus many branched chain and cyclic materials

The efficiency of collection of the most volatile gases by molecular sieves can be improved by cryogenic cooling (Table 5.12).

Table 5.12 Low temperature purification of gases on 5A molecular sieves (Zief, 1976)

Gas	Coolant	Impurities removed
Hydrogen[a]	N_2(liq)	N_2, Ar
Helium	N_2(liq)	N_2, Ar, O_2
Nitrogen	O_2(liq)[b]	CO, O_2
Argon	O_2(liq)[b]	N_2, CH_4
Oxygen	CO_2(s) slush	C_2H_6, CO_2, Kr

[a] Oxygen must be removed before use to prevent a dangerous build-up of this gas.
[b] Care should be taken to exclude organic materials which can be dangerously oxidized by liquid oxygen.

Catalytic convertors can be employed to change contaminants into another form, the conversion being either to a compound which does not interfere in the analysis or to one which can be more readily collected by an adsorbent or similar collection system. Some examples of catalytic systems are as follows.

- The removal of alkane impurities from an inert gas by combustion over copper(II) oxide. The alkanes are oxidized to carbon dioxide, which can then be stripped from the gas stream by cryogenic trapping or with molecular sieves.
- Hydrogen can be removed from inert gases by passing the gas over copper(II) oxide heated to 225–280 °C. The hydrogen is converted to water, which can then be collected using a suitable drying agent.

REFERENCES

BDH Ltd, (1984) *AnalaR Standards for Laboratory Chemicals,* 8th Ed. BDH Ltd, Poole, Dorset, UK.

Bio-Rad Laboratories. (1982) *Ion-exchange Manual,* Richmond, CA.

Falbe, J. and Hahn, H.-D. (1974) In *Methodicum Chimicum,* F.Korte (Ed). Academic Press, New York, Vol 1(B), p.990.

Health and Safety Executive (1992) EH40/92. HMSO Publishers, London.

Korkisch, J. (1989) *Handbook of Ion-exchange Resins: Their Application in Inorganic Analytical Chemistry,* Vol 1. CRC Press, Boca Raton, USA.

Kuehner, K., Alverez, R., Paulsen, P.J. and Murphy, T.J. (1972) *Anal. Chem.* **44**, 2050.

Mattinson, J.M. (1972) *Anal. Chem.* **44** (9), 1715.

Perrin, D.D and Armarego W.L.F. (1988) *Purification of Laboratory Chemicals.* Pergamon Press, Oxford.

Zief, M. (1976) *Contamination Control in Trace Element Analysis.* Wiley, New York.

6 The Water Supply

The most basic requirement of any analytical laboratory is a good water supply. However, trace analysis laboratories, often carry out a variety of analytical procedures, each of which requires a different grade of water. There is no one water supply which is appropriate to all applications and it is important to identify the criteria which have to be applied in the selection of a suitable purification system. In many procedures the major criterion is to obtain a suitably low reagent blank. For the analysis of dissolved zinc, for example, the first priority must be to choose a purification method to minimize the zinc content of the water. It is fairly obvious to state that the materials used in the construction of the purification apparatus should be free from zinc, but identifying plastics which have not been manufactured using zinc compounds may not be as easy. This particular aspect of the materials which are used in the construction of laboratory apparatus is dealt with elsewhere in the book. Contamination arises not only from the materials which are employed in the construction of the purification system but also from external sources such as atmospheric fallout. Minimizing the blank may not be the only criterion which is important; water used in the measurement of reduced species, for example, may require particular attention to be paid to its dissolved oxygen content.

There are many proprietary water purification systems available, based on a number of different operating principles. Each has its own set of advantages and disadvantages, and it is normal for more than one purification method to be combined to obtain the purest product.

6.1 GENERAL MEASURES OF WATER QUALITY

The final decision as to whether the quality of a water stock is suitable for use in a particular trace analysis can only be made from its performance. It is, however, unlikely that a water which is high in dissolved material will be useful for many analyses, and general measures of quality will give a reasonable starting place for choosing which source to employ in a particular application.

In most laboratories the purchase of pre-purified water is not cost effective and raw water is normally therefore purified on site. The normal source of water is the domestic water supply, which varies greatly in composition from one area to another. Whilst there is no one parameter which on its own is capable of assessing the quality of a water, a number of measurements together can be usefully employed to indicate the overall quality of a water source and the effectiveness of purification procedures. When considering the purchase of water purification systems it is important to be able to appraise the performance of a system from published literature. A number of performance parameters are often quoted for such apparatus; as with the parameters which are used to assess the quality of the untreated water, none gives a complete description of the water quality, but together they are useful indicators.

6.1.1 Water Hardness

Of prime concern when selecting suitable plant to be used in the construction of a new laboratory water supply is the water hardness. This is a measure of the propensity of the water to deposit a scale, and is essentially a measure of the concentration of calcium and magnesium ions in the water. The total hardness of the water can be defined as the sum of the concentrations of these two species in the water. The calcium and magnesium can be derived from a number of mineral sources resulting in their presence as bicarbonates, carbonates, sulphates and chlorides. Which of these salts is present determines whether the water is described as having 'temporary' or 'permanent' hardness. When the water is boiled the bicarbonates decompose to carbonates of lower solubility, which then precipitate out of solution. This results in a reduction of the salt concentration in the water and this hardness is described as being 'temporary'. The chloride and sulphate salts on the other hand are unaffected by boiling and are therefore described as 'permanent'.

6.1.2 Total Dissolved Solids (TDS)

This is the weight of all of the involatile organic and inorganic solutes in a litre of the water. It is expressed as parts per million or mg/l. The common extremes range from 35 000 ppm in seawater to less than 1 ppm in purified waters. As a measure of water purity it is most usefully applied in the appraisal of the hardness of water supplies which are to be purified.

6.1.3 Conductivity and Resistivity

A general indication of water purity can be obtained by the measurement of its conductivity. This is the most convenient means of monitoring changes in system performance with time, and a simple conductivity measuring device is often incorporated in the design of water purification systems. Conductivity is only a gross indication of the dissolved ionic impurities in the water and cannot be taken as a guarantee of purity for a particular application. In particular many impurities, most notably organics and particulate matter such as bacteria, conduct poorly, if at all, and do not therefore contribute to the reading.

Measurements of conductivity can be carried out either using an in-line cell or

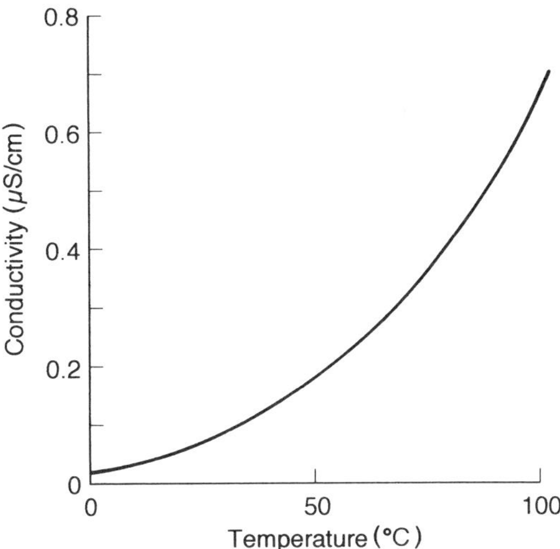

Figure 6.1 The variation of the conductivity of ultra-pure water with temperature

with hand or bench apparatus. The range of conductivity to be expected for source and purified waters is from 500 μS/cm (50 mS/m or 0.002 megohm.cm) for hard mains water to 0.05 μS/cm (0.005 mS/m or 20 megohm.cm) for a very high purity water. However, conductivity is temperature dependent and increases by between 2 and 7% per °C depending on the purity of the water (Figure 6.1).

6.1.4 pH

Ideally, water has a pH of 7, but high purity water has no significant buffering capacity and its pH can swing violently if additional material such as atmospheric carbon dioxide is dissolved in it. Carbon dioxide dissolution is a particular problem encountered with water purified by distillation. It should be pointed out that it is very difficult to measure the pH of high purity water with conventional pH electrode systems. In addition, water contamination is inevitable with conventional electrodes, as they bleed silver or mercury salt contaminated potassium chloride solution continuously from the reference electrode.

6.1.4.1 Buffer solutions for pH measurement

The potential/pH response characteristics of an electrode combination vary throughout its working life. Regular accurate calibration of the measurement system is therefore necessary; under unfavourable circumstances this may be hourly. Of the many buffer systems which have been devised for the calibration of pH meters the following solutions have been widely accepted.

To prepare buffer solutions having nominal pH values of 4, 7 and 9, dissolve the following in distilled water (boiled for 15 minutes and then cooled to room temperature) and make up to 1 dm^3. The pH values are quoted at 20 °C.

pH 4.00 10.21g of anhydrous potassium biphthalate ($KHC_8H_4O_4$)

pH 6.88 3.40 g of anhydrous potassium dihydrogen phosphate (KH_2PO_4) plus 3.55 g of anhydrous disodium hydrogen phosphate (Na_2HPO_4). Both reagents should be dried at 120 °C for 2 hours before use.

pH 9.23 3.81 g of sodium borate decahydrate ($Na_2B_4O_7.10H_2O$)

The pH of these buffer solutions is dependent on temperature (Table 6.1).

Table 6.1 The temperature dependence of pH buffer solutions

°C	'pH 4'	'pH 7'	'pH 9'
0	4.00	6.98	9.46
5	4.00	6.95	9.40
10	4.00	6.92	9.33
15	4.00	6.90	9.28
20	4.00	6.88	9.23
25	4.01	6.86	9.18
30	4.02	6.85	9.14

Buffer solutions should be prepared regularly and stored, cool and dark, in polyethylene containers to avoid algal and mould growth and contamination.

6.4.1.2 Setting up the pH meter

A pH meter measures the potential of a sensing electrode relative to that of a reference electrode. The device is therefore a disguised voltmeter which, instead of volts, displays pH. There is normally a linear relationship between the measured cell potential and the pH of the solution; setting up the pH meter involves adjusting the displayed pH to the correct value to take into account that electrodes differ slightly in their response characteristics.

There are normally only two main controls on the meter; these are:

(1) Intercept, Set buffer, Asymmetry or Standardize;

(2) Slope, Temperature or Offset.

Modern electrodes are normally supplied to give a zero cell potential when placed in a pH 7 buffer; this is sometimes called the zero-point pH or isopotential point and is often written as $E°7$.

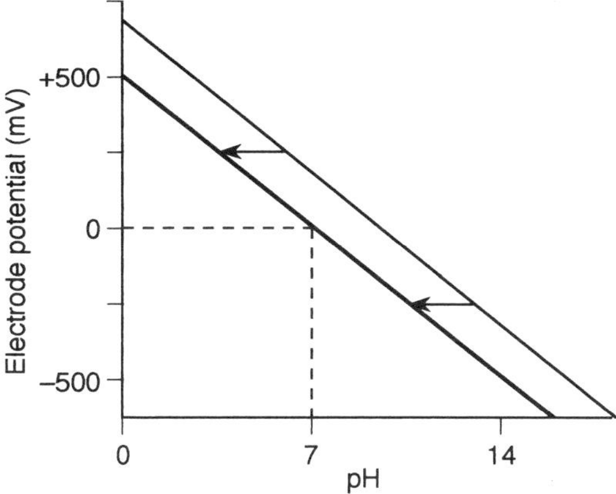

Figure 6.2 Setting the Intercept control of a pH meter

The first step in setting up the system is to place the electrodes in a pH 7 buffer and to use the Intercept control to shift the response curve sideways until the meter reads the correct pH (Figure 6.2).

The electrodes are then placed in another buffer having a pH which, together with the pH 7 buffer, brackets the range over which the electrode is to be used. The Slope control is then adjusted until the meter displays the correct pH of the second buffer. This results in the response curve being rotated around the isopotential point until the correct pH is displayed (Figure 6.3).

The meter should now be ready to use but it may be necessary to repeat the adjustments to obtain the most accurate calibration.

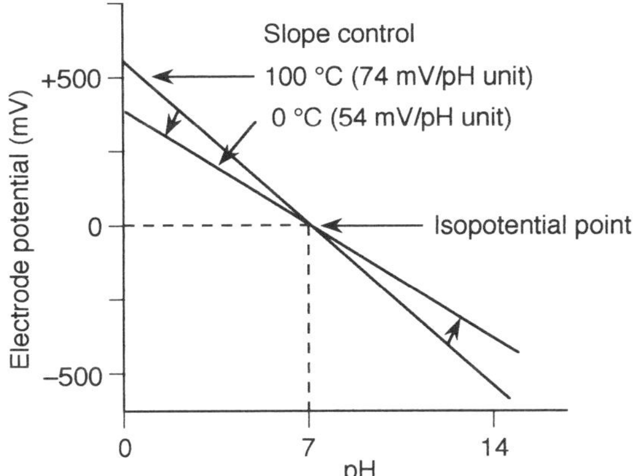

Figure 6.3 Setting the Slope control of a pH meter

6.1.5 Dissolved Organic Carbon

Dissolved organic carbon (DOC) measurements are of particular use when it is necessary to ensure that the quantity of organic matter either remaining in the water after treatment, or leached from the materials employed in the purification, are low. There are a number of ways that this can be carried out but none is either straightforward or readily available in the conventional laboratory. Instrumentation for DOC analysis is generally based on the oxidation of the organic carbon to carbon dioxide, and the resulting gas is then measured by either gas chromatography with a thermal conductivity detector or by infrared absorption.

6.1.6 The Quality of the Domestic Water Supply

Whilst most domestic water supplies have undergone some degree of purification before they reach the laboratory, they will still largely reflect the geology of the region from which they are drawn and the additives which have been used in their treatment (Table 6.2).

Table 6.2 Water supply analyses for four different areas in the UK

Parameter	Area 1	Area 2	Area 3	Area 4
pH	8.29	8.64	7.72	7.36
Conductivity (μS/cm)	75	72.4	434	522
Hardness (mg $CaCO_3$/l)	12.8	23.3	179	263
Alkalinity (mg $CaCO_3$/l)	9	14.3	211	233
TOC (mg C/l)	2.5	1.23	2	1.6
Chlorine (mg/l)	<0.05	0.19	0.06	0.25
Chloride (mg/l)	11	6.29	29	14.7
Nitrate (mg $NO_{3/}$l)	2	0.76	22	24.8
Nitrite (μg NO_2/l)	4	10	7	<30
Phosphate (μg P/l)	24	<10	58	<40
Sulphate (mg SO_4/l)	17	8.29	30	7.7
Fluoride (μg F/l)	82	<100	120	110
Boron (μg B/l)	<50	<200	99	<200
Sodium (mg Na/l)	5	3.6	17	8.5
Potassium (mg K/l)	0.6	0.4	3	1.2
Calcium (mg Ca/l)	11	7.05	73	104
Magnesium (mg Mg/l)	2	1.25	3	2
Barium (μg Ba/l)	14	<200	35	17
Iron (μg Fe/l)	36	64.3	55	9.8
Aluminium (μg Al/l)	48	88.5	19	13.5
Copper (μg Cu/l)	26	32.6	330	140
Zinc (μg Zn/l)	10	31.4	16	22
Nickel (μg Ni/l)	<5	<5	3	<5
Arsenic (μg As/l)	<0.2	<5	0.8	<0.2
Manganese (μg Mn/l)	6	8.83	5	<10
Silver (μg Ag/l)	<1	<1.0	0.2	<2
Lead (μg Pb/l)	9	3.94	26	5.1

The complexity of the task of water purification, and the nature of the plant which is best suited to the task, will as a consequence depend on the nature of the raw feedstock.

6.2 METHODS OF WATER PURIFICATION

6.2.1 Distillation Methods

Distillation is an effective means of water purification but it is a costly consumer of electricity and cooling water. If the still is fed directly from the mains in hard water areas, the formation of deposits will result in a drop in efficiency with time, and frequent descaling will be necessary.

The materials used in the construction of the distillation apparatus have important consequences in the quality of product. There is little to be achieved by using high purity materials for the construction of stills which are used for the preliminary purification of a raw feed water but, for the 'polishing' of water for low level trace metal analysis, for example, greater thought needs to be given to the construction. For trace metal work the still should be constructed from materials such as silica and PTFE and submerged metallic heating elements must be avoided.

6.2.2.1 Conventional distillation

Conventional distillation by boiling significantly reduces the contaminant load of a raw feedstock, but the level of purity which can be attained is highly dependent on still design and operation. Aerosol formation by the boiling process leads to droplets of unpurified material being carried over with the distillate, and the purity of the product will therefore depend on the build-up of material in the boiler chamber. To minimize this effect, the boiling should be as smooth as possible and an efficient trap for droplets should be provided. The path between boiler and receiver must therefore be made tortuous and/or long. This can be achieved by the use of a distillation column filled with silica helices, but this will normally result in a reduced yield.

Since the quality of the product from a still depends on the quality of the material in the boiler, the use of two stills in series significantly improves the quality of the final distillate.

6.2.1.2 Sub-boiling distillation

The problem of droplet carry-over associated with the conventional still is overcome by maintaining the temperature of the feedstock below its boiling point. Heating the surface of the liquid in a closed container increases the vapour pressure of the reagent, and this vapour can then be recondensed. As there is no boiling step, there is no droplet formation and contaminant carry-over is reduced. Typical sub-boiling distillation units have been described in the previous chapter on reagent purification.

6.2.2 Reverse Osmosis

Reverse osmosis involves the separation of dissolved material from a solvent by the application of pressure to force raw water through a membrane, against osmotic pressure. The membrane preferentially allows the water molecules to pass through, rejecting a large proportion of the dissolved (and particulate) contaminants (Table 6.3). The main advantages of the technique are that it can be employed in a continuous process requiring little attention and it is not energy intensive. In the laboratory it is finding increasing use as a preliminary purification method, especially in areas where the tap water contains high levels of total dissolved solids (TDS).

Table 6.3 The rejection of water impurities by reverse osmosis

Component	Rejection
Monovalent ions	>95%
Polyvalent ions	>97%
Particles	>99%
Bacteria	>99%
Organics >300MW	>99%

6.2.3 Ion-exchange

Water softening plant and water purifiers are related systems employing ion-exchange resins. The two processes should not, however, be confused as water softeners are normally only designed to replace the troublesome elements of hard water by sodium chloride. Although this reduces scale build-up if the water is subsequently distilled, the process has not reduced the concentration of dissolved material in the water. In a water purification system, however, percolation of raw water through a bed of mixed anion (hydroxyl form) and cation (hydrogen form) exchange resins swaps charged contaminants for hydrogen and hydroxyl ions, resulting in the substitution of the contaminant salts by water.

$$\text{RESIN—H}^+ + \text{impurity cation} \rightleftharpoons \text{RESIN—(impurity cation)} + \text{H}^+$$
$$\text{RESIN—OH}^- + \text{impurity anion} \rightleftharpoons \text{RESIN—(impurity anion)} + \text{OH}^-$$

$$\text{H}^+ + \text{OH}^- \rightleftharpoons \text{H}_2\text{O}$$

The technique of ion-exchange is discussed in greater detail elsewhere in the book. It is an effective means of removing a wide range of charged contaminants (Table 6.4), and is relatively inexpensive to install and simple to operate. Its main drawback is that it can be rather expensive to operate in hard water areas. Unless only small quantities of water are required, the technique is therefore best employed as a secondary purification method following, for example, reverse osmosis.

Table 6.4 Typical performance parameters obtained from the purification of water by deionization

Parameter	Tap water	Deionized water
Conductivity	250 μS/cm	2–30 μS/cm
Resistivity	0.0004 mΩ-cm	0.03–5 mΩ-cm
TDS	200 ppm	<10 ppm
pH	6.5–8.5	4–8
Heavy metals	<100 ppb	<5 ppb
Silica	2 ppm	2 ppm
Dissolved CO_2	<30 ppm	present

In order to achieve the purification of water it is necessary to employ both anion- and cation- exchange systems. Although it is possible to employ cation- and anion-exchange resin beds in series, leakage of ions from each resin, due to the reversible nature of the exchange reaction, results in the water containing residual salts. This problem can be most effectively overcome by mixing the cation- and anion-exchangers together to give a mixed-bed system.

In hard water areas the resin lifetime is short and frequent regeneration may be necessary. To estimate the maximum potential lifetime of such a system in a particular area, it is possible to base calculations on an average capacity of 5 milli-equivalents (for divalent ions this is 2.5 mmol) per gram of dry resin.

Once ion-exchange resins have reached their capacity they can be regenerated; the anion-exchange resin with dilute sodium hydroxide and the cation-exchanger with dilute hydrochloric acid. The situation is more tricky with mixed-bed systems, as the two resins are intimately mixed and cannot be regenerated with a common reagent. In an emergency the anion and cation resins can normally be separated by differential flotation in saturated brine; they can then be separately regenerated, as appropriate, with acid or base. After rinsing with water the resins can then be re-mixed and returned to the deionizer cartridge. This process is labour intensive and it is therefore recommended that the cartridges be returned to the manufacturer for regeneration, or more commonly nowadays for replacement of the resin.

As they are made of a plastic (styrene—divinylbenzene copolymers are typically used), the ion-exchange resins are prone to release trace organic matter into the water. This is particularly so with the anion-exchanger, which is normally a quaternary ammonium modified polymer. Each time the system is restarted water should be discarded until the conductivity of the output drops to a steady acceptable value.

6.2.4 Multi-stage Purification

For the preparation of high quality water at least two steps are generally necessary. Preliminary purification by reverse osmosis, distillation or ion-exchange is followed by a second step which may be, for example, a second distillation or further ion-exchange. As with all trace analysis problems, there is no universally applicable source of pure water, and the ultimate test is the measurement of the analyte in the water and its contribution to the reagent blank. For general purpose use, however, it is normally better to employ two stages which work on different principles than to duplicate one purification technique. When employing two methods in series, the sequence in which they are carried out can be important, as although the second stage is employed to remove residual impurities from the water, it may in turn introduce its own contaminants.

6.3 SUMMARY

There is no ideal water supply for all applications. In addition to the overall assessment of water quality, which can be made by the measurement of such parameters as conductivity, additional criteria such as the pH, dissolved oxygen content or residual organic matter may be of significance for a particular application. The most important criterion is, as ever, the suitability of the water for the analytical procedure. The performance of various systems will vary according to the quality of the feed water, initial system design, age, and state of maintenance; these are difficult to predict. Table 6.5 gives a general summary of the conductivities which might be expected from a variety of water purification systems.

Table 6.5 Performance of water purification methods

Water type	Conductivity μS/cm	Resistivity MΩ.cm
Hard mains	500	0.002
Medium hard mains	200	0.005
Soft mains	75	0.015
Reverse osmosis (RO)	10–75	0.1–0.013
Distillation	1–5	1–0.2
Deionization	0.1	10
RO/nuclear grade deionization resin	0.05	18

7 Working Practices

7.1 INTRODUCTION

This section considers the analytical steps which occur between the collection of a sample from a production line or in the field, up to and including the final determination step. It is not intended to provide detailed descriptions of different analytical techniques, but to focus on the general approaches used in such analytical sequences and to indicate the strategy needed to ensure the generation of good data. The use of carefully selected and prepared reagents and equipment and a well chosen environment for the work, as described earlier, are assumed as the backdrop to the procedures described here. In analytical papers and monographs the curt style which is often used means that superficially 'minor points' are missed or considered as falling under the heading of 'good laboratory technique'. If a thinking and critical approach to the analysis is not adopted, particularly at very low analyte concentrations, a crucial step can invalidate the whole measurement. At all stages the general objective is to ensure that data of an acceptable quality are produced. Successful trace analyses are based on carefully designed and executed protocols, which in turn are based upon a knowledge of potential problems which can arise during the work.

7.2 SAMPLING

Although detailed sampling strategy is beyond the remit of this book, if accurate data are to be obtained sampling cannot be isolated from the whole measurement sequence.

> The chain of operations is only as good as the weakest link

Sampling may be as simple as picking a unit from a production line or as difficult as bringing back geological samples from the moon. However, some specific factors are particularly important in trace analysis.

The sampling device is clearly a major potential source of contamination or loss of analyte. Modification may be required to remove components from commercial apparatus which can release inorganic contaminants into the sample. It has been necessary, for example, to replace conventional rubber seals and 'O' rings in sampling bottles used for the collection of seawater for trace metal analysis with ones made of silicone rubber. The conventional plastic tap was also replaced by one made of PTFE. The means of deploying and activating such samplers cannot be ignored. A plastic line is now used to deploy the samplers, as conventional metallic cables have been shown to be a source of contaminating particulate iron. Similar problems with contamination can occur when sampling ice cores. Here the major contamination potential is from the coring device. As contamination of the exterior of the core material is virtually impossible to prevent with conventional coring techniques, and as the sample is a solid, the approach has been to remove and analyse successive concentric layers. When sequential analyses give the same result, it is assumed that no significant contamination is present (Wolff, 1990). In clinical analysis of trace metals in body fluids, trace-metal-free syringes for blood collection are essential, and acid cleaned or previously checked batches of bottles should be used, for example, for urine collection. The storage of samples, once collected, is considered in more detail in a separate chapter.

7.2.1 Cleaning and Processing of Samples

If a specific fraction of the overall sample is required, it may be necessary very carefully to clean a sample prior to analysis. In palaeo-oceanographic studies, the ratio of cadmium to calcium in benthic foraminifera can be used as an indicator of nutrient concentrations in palaeo oceans where cadmium acts as a proxy for phosphate. The cadmium in the calcite tests of the organism must be measured, but these tests can be coated with manganese oxides, which also remove some cadmium from seawater. A very careful clean-up stage to remove the coating is therefore required if a meaningful measurement is to be made. A further example is the potential for the gut contents of organisms to bias the analysis of the tissue.

Another operation which has sometimes been undertaken prior to analysis by surface probe techniques is staining of tissue samples. Any such processing must be done with caution. An example lies in the link between aluminium and Alzheimer's disease, where the most direct evidence for this link is based on the presence of aluminium in specific brain tissues. This earlier work now needs to be reassessed, as research using nuclear microscopy techniques suggests that much of the aluminium observed in earlier work may have originated from aluminosilicate contaminants in the stain which was used (Landsberg et al., 1992).

7.3 GENERAL PRINCIPLES OF HANDLING SAMPLES

Assuming a clean sample has been taken, the next potential problem areas are transport, storage, and handling in the laboratory. Again it is important to identify the critical steps in which contamination and losses may occur. Whilst systematic contamination can be introduced at this stage, through, for example, poorly cleaned equipment, blank problems can just as readily occur as random events. The steps which are necessary to reduce this latter type of problem during general handling steps are considered here first. Some of the actions may at first sight seem facile, but experience has shown that the precautions become increasingly important as the concentration of analyte becomes lower.

Each additional stage in an analysis can introduce contamination, and minimising the number of handling steps is therefore an important principle.

Minimize the numbers of containers and solution transfers in all procedures to reduce contamination

Analytical techniques which directly measure the analyte are therefore advantageous, provided that the measurement system itself is non-contaminating. An example of a potentially redundant handling procedure is the quantitative transfer of an aqueous sample with a measuring cylinder. An alternative is to transfer the sample directly to a tared container and then to reweigh. Reagent solutions can also be made directly in their storage containers, using pre-calibrated bottles and gravimetric additions.

If a sample is in a clean capped reservoir, then the potential for a contaminant to enter that container should be negligible. That is, unless one is dealing with volatile analytes, such as mercury, which are capable of diffusing through plastics. Once the bottle is opened, however, contamination can occur through the entrainment of atmospheric particles passing across the bottle mouth, or by the direct entry of larger particles. There are several obvious steps to reduce these problems; only open bottles for the shortest period possible, and then only in a clean environment. Avoid passing the arms or other items directly across open sample bottles, and ensure that, when the top is separate from the bottle, it is placed in a location where it will not pick up contamination. When pouring samples, ensure that the lip of the mouth of the container does not contact the reservoir or other potentially contaminating items.

Whilst in practice solid phase samples often contain higher concentrations of analyte than solutions, there can still be significant potential for contamination during sub-sampling. Obviously processing equipment containing the analyte or potentially interfering material, such as metal sieves, should be avoided, and the use of an acrylic spatula or similar plastic tool for sub-sampling should be mandatory. A representative sub-sample of a powdered sample can be obtained by forming a cone of the material and quartering on a cleaned PTFE sheet, using plastic tools.

Within a clean working area there is usually a gradient in the cleanliness of the working environment. In laminar flow hoods the area closest to the HEPA filter

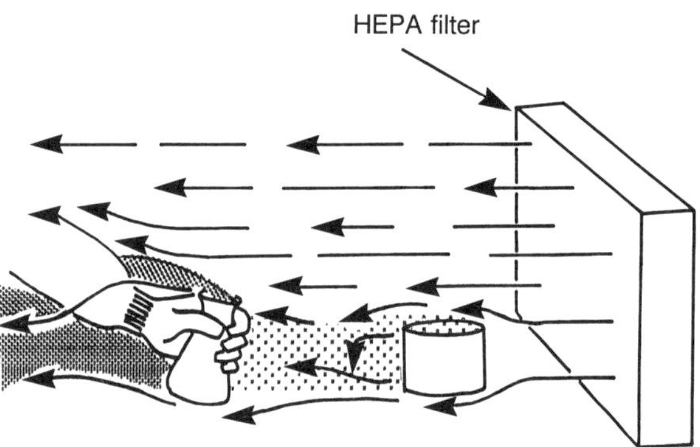

Figure 7.1 Increasing contamination potential down-stream of a filtered air supply

will be the cleanest. Down-stream of any equipment or other item in the hood, there is the potential to pick up particles generated by the item (Figure 7.1). It is necessary to be aware of this type of gradient when planning laboratory activities, and to focus the most critical activities in the cleanest locations.

Such practices are only examples of the approaches which can be taken in the clean handling of samples. It is essential to think laterally, and for analysts to modify and extend the principles to their own particular circumstances. There are two general ways of approaching an untested trace analysis. One can try a basic approach where limited clean procedures are incorporated, or one can move to more rigorous methods from the start. Although the former may be appropriate for a particular application and may involve less resources, in less well controlled situations regular blank measurements are essential to identify the random problems that may arise (see Chapter 8, Trouble-shooting). The point at which more sophisticated approaches have to be taken depends on experience of the system. As a general rule, the more rigorous approaches should be the norm in new areas where suitable standards may not be available and the accurate sample concentration is not known at the outset. Extreme contamination control measures may be necessary in a routine environment into which new unskilled staff are regularly recruited. Some relaxation of procedures may be appropriate when the ability to measure accurately the concentrations of analyte has been demonstrated, and if staff are sufficiently skilled at this point to take the necessary precautions.

7.4 FILTRATION

Filtration is an important and common procedure in many areas of trace analysis where the separation of particles from gases and fluids is required. Applications include the collection and isolation of samples for industrial hygiene and geochemical studies and the sampling of atmospheric particles. At the analysis stage, filtration is employed for the collection of precipitates and in the removal of contaminating particles from gases which are to be used in critical applications such as the stirring of electroanalytical cells.

There are several important reasons why sample solutions often are filtered prior to their analysis. Valuable information may be obtained from the composition of the particles, and the particles may represent a significant fraction of the total element present. If the total sample is to be acidified, e.g. for storage, then a fraction of the element associated with the particles can be released into solution and will be wrongly considered as a part of the dissolved phase. The dissolved phase concentration may also be reduced if the particles are reactive to the dissolved form of the element and remove them from solution. Such particles may be either inanimate adsorptive solids or organisms such as bacteria. Analytical factors can also be important; high concentrations of solids can block nebulizers in atomic spectrometry instruments and scatter light in colorimetry. An ideal filter should (as modified from Riley *et al.*, 1975):

(1) have a uniform and reproducible pore size;

(2) not clog easily;

(3) if necessary retain particles at the surface to aid analysis by microscopy and its associated spectroscopic techniques (such as SEM-EDAX);

(4) readily equilibrate with the surrounding atmosphere during gravimetric determinations;

(5) not adsorb the trace elements to be analysed, or species containing them;

(6) not contain significant amounts of the elements to be determined;

(7) have reasonable mechanical strength;

(8) not shed fibres or otherwise disintegrate.

Filters fall into two main types as follows.

(1) *Depth filters* are formed from a randomly distributed matrix of fibres or small particles. There is no well defined pore size, and the separation relies on physical trapping and surface contact. Generally the thicker the filter, the smaller the effective size of particle which passes through. The matrix can be made from a variety of materials including paper, metal oxide precipitates and glass fibres. Metal oxide filters rarely find applications in trace analysis because of their contamination potential for many elements. The conventional paper filter can also sometimes give

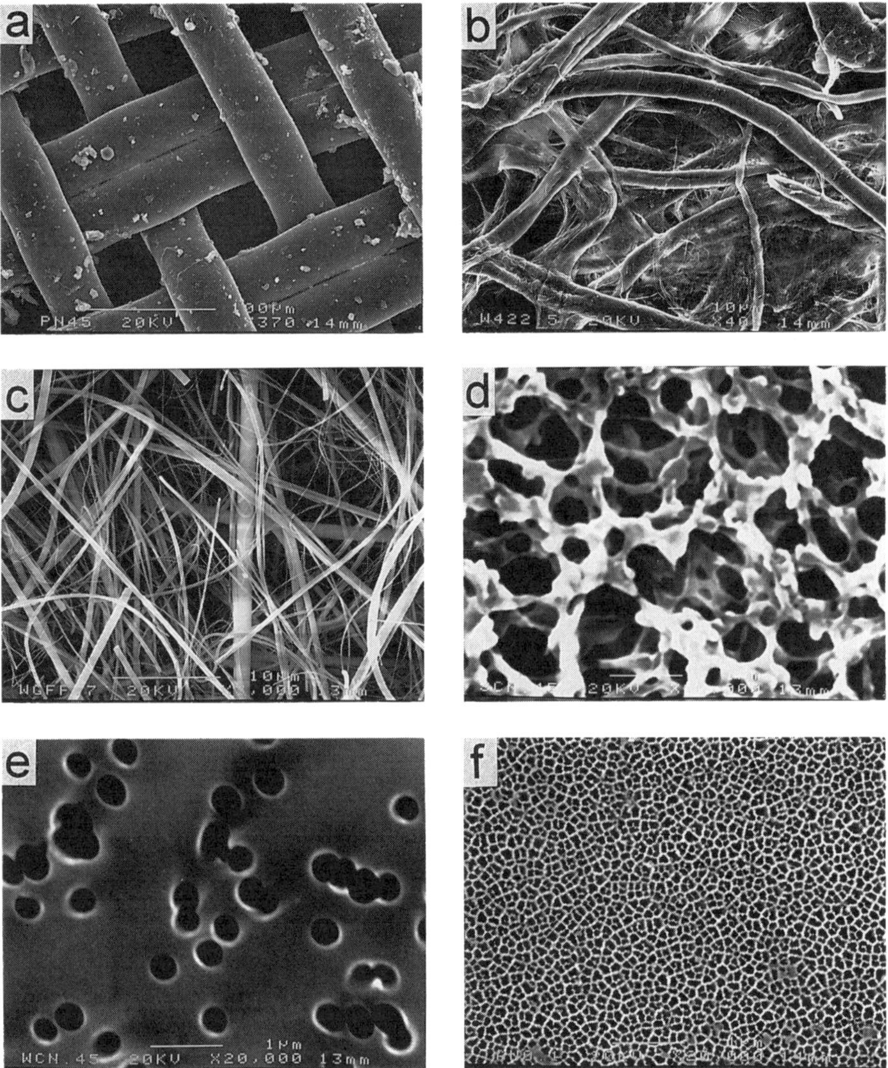

Figure 7.2 Scanning electron micrographs of commonly used filtration materials. The nominal diameter of particles retained on these materials is given in brackets, together with the magnification used: (a) phytoplankton netting (45 µm, × 185); (b) paper filter (Whatman 42; 2.5µm, × 200); (c) glass fibre filter (Whatman GF/F; 0.7 µm, × 200); (d) cellulose nitrate membrane (Sartorius, 0.45 µm, × 10 000); (e) polycarbonate membrane (Nuclepore; 0.4 µm × 10 000); (f) aluminium oxide filter (Whatman Anopore; 0.1 µm, × 10 000)

problems with adsorption and, for some elements, contamination. The most commonly used analytical filter of this type is made of glass fibres. A high purity (and-cost) version manufactured from silica fibres is useful for the measurement of total element content by neutron activation analysis (NAA), where activated sodium in glass filters can be a problem.

(2) *Barrier filters* have much better defined pore sizes, and are very popular in trace analysis because of their purity. They are made principally from plastic films of cellulose ester or polycarbonate. Some types have very well defined holes, obtained by etching small holes to the appropriate size. The filters are inherently clean for many elements, and relatively easy to purify, but the quantity of material which can be collected on these filters is normally low. It is important to realise that, as with other types, the effective pore size will change as more and more particles are collected onto the filter.

One type of filter has a honeycomb-like structure of aluminium oxide, with well defined channels extending through the medium. As the walls of the channels are thin, the filter has a high filtration capacity, with effective pore sizes down to 0.02 μm. Clearly the filter cannot be used for aluminium determinations, but this design may prove to be appropriate for other trace analytes.

The structure of commonly used filter media is shown in Figure 7.2.

The final choice of filter is a compromise between the various requirements of the analysis. There is no fixed demarcation line between the dissolved and particulate states, and the filter size cut-off which is employed to separate particulate from dissolved material varies from one field to another. For most studies of transition metals in natural waters, 0.4 μm has been taken as the arbitrary cut-off. Other factors influencing the choice of filter are compatibility with subsequent analytical steps and contamination potential. Thus, while polycarbonate membranes have been used in many natural water investigations, some appear to suffer from chromium contamination. These membranes are also unsuitable for the study of mercury because of adsorption and contamination problems, and 0.7μm mean pore size glass fibre filters are used instead. The cleaning of filters depends on the analyte and the material of the filter. Most procedures use techniques which have been described elsewhere in this book for the decontamination of apparatus. Obvious potential for contamination should be avoided. Stating what should by now be obvious, nitric acid, for example, should not be used as a cleaning agent if nitrate is to be determined in the filtrate. It is extremely difficult to remove acid totally, particularly as it can diffuse into many plastics in aqueous and dissociated nitrogen oxides forms, and the filtrate will become contaminated with the anion. Where the analytes are volatile and the filter medium is stable at elevated temperatures, heating can be used. This is particularly suitable for the treatment of glass fibre filters for Hg and P analysis. It is always essential to test the blanks of filters and to assess losses due to sorption onto the filter.

The filtration operation can be done in several ways, as described below.

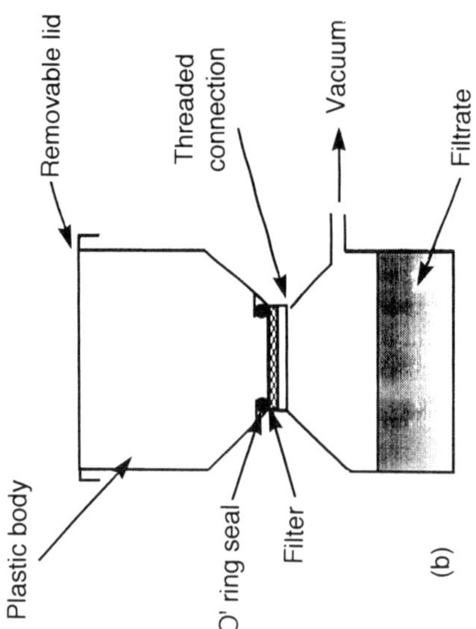
Removable lid
Threaded
connection
Vacuum
Filtrate
Plastic body
'O' ring seal
Filter
(b)

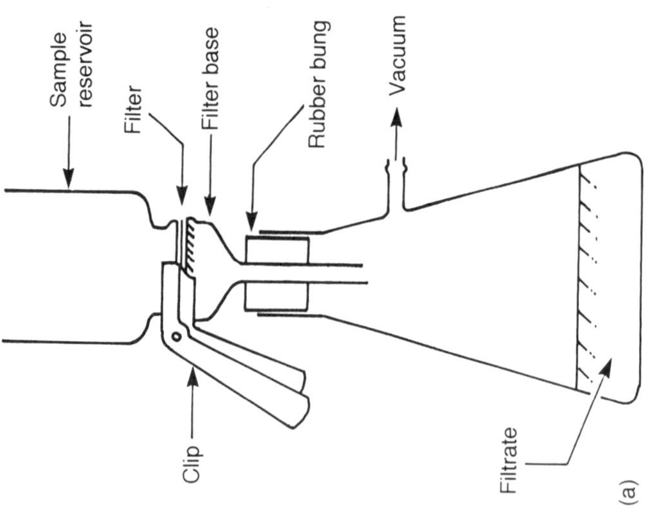
Sample
reservoir
Filter
Filter base
Rubber bung
Vacuum
Clip
Filtrate
(a)

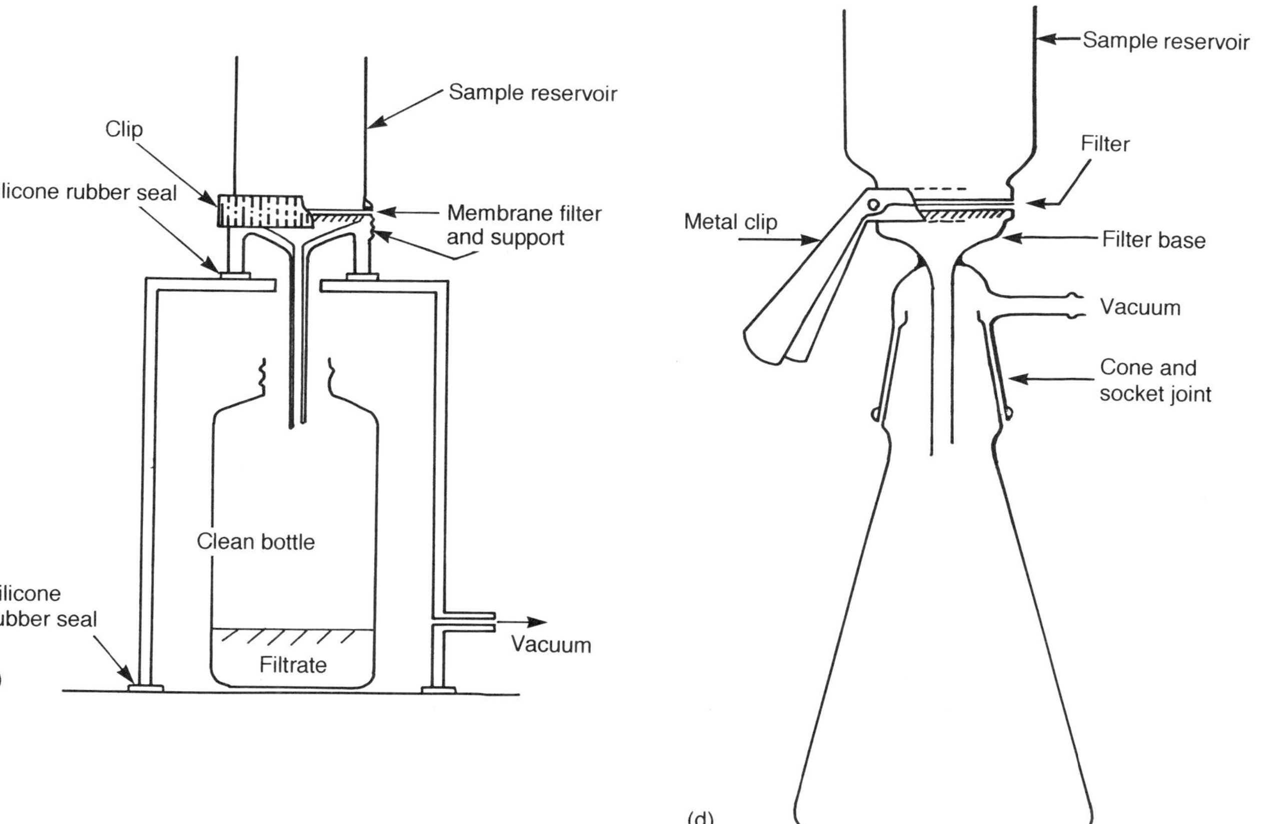

Figure 7.3 Different apparatus for the vacuum filtration of aqueous samples: (a) conventional filter flask assembly; (b) commercial all-plastic filtration system; (c) use of a vacuum jar for the clean filtration of samples; (d) glass vacuum flask, filter support and reservoir, designed to minimize contamination of the filtrate

7.4.1 Vacuum

Vacuum filtration has inherent potential for contamination because of the nature of the transfer steps and the number of them. The traditional equipment of filter holder, bung and vacuum flask [Figure 7.3(a)] has several disadvantages. The bung can be a major source of contamination, particularly as the filtered solution must be poured past the lip which has been in contact with the bung. If a metal clip is used to hold the sample reservoir and base together, any spillages can become contaminated on running down past this onto the bung and lip of the flask.

A better system, which is made entirely of plastic and does not use the conventional bung, is shown in Figure 7.3(b). The sample must still be transferred to the next container. An improved approach is to use a vacuum jar as shown in Figure 7.3(c). In this arrangement the sample passes directly from the filter holder and reservoir into the clean receiving container, and no further transfer is needed. For most trace elements, the reservoir for the filter unit should be of plastic with a non-contaminating 'O' ring or sealing surface.

Where no contamination or loss problems occur with glass, or where plastic materials can sorb or be a source of analyte (such as with mercury), borosilicate equipment can be used. The preferred design has a cone on the vacuum flask [Figure 7.3(d)], to prevent contaminating spillages entering the filtrate via the joint and to give a direct path for the filtrate from the reservoir without contacting the joint.

7.4.2 Pressure

Pressure filtration has several advantages over vacuum filtration. It is more readily adapted to in-line filtration and the sampling container itself can be employed as the sample reservoir. This process minimizes the number of transfer steps and the sample is not exposed at any stage during the filtration. Water sampling bottles, for example, can often be pressurized by filtered nitrogen, to force the sample through an in-line filter directly into a clean sample bottle. A range of in-line filter holders is available commercially. The general comments made earlier about the suitability of different materials also apply here; non-pigmented materials are generally best. In-line filtration apparatus is not without problems, some examples of which include:

(1) contamination from 'O' rings made from butyl rubber and similar materials; FEP encapsulated 'O' rings are a good alternative;

(2) air locks in the cavity immediately above the filter can stop or slow filtration and be difficult to remove. Holders incorporating purge valves overcome this problem.

A design for a 47 mm diameter filter holder made from PTFE and incorporating these features is described by Morley et al. (1988), and is shown in Figure 7.4(a). Extended leaching of the unit with a clean sample solution will indicate if there is any release of contaminant.

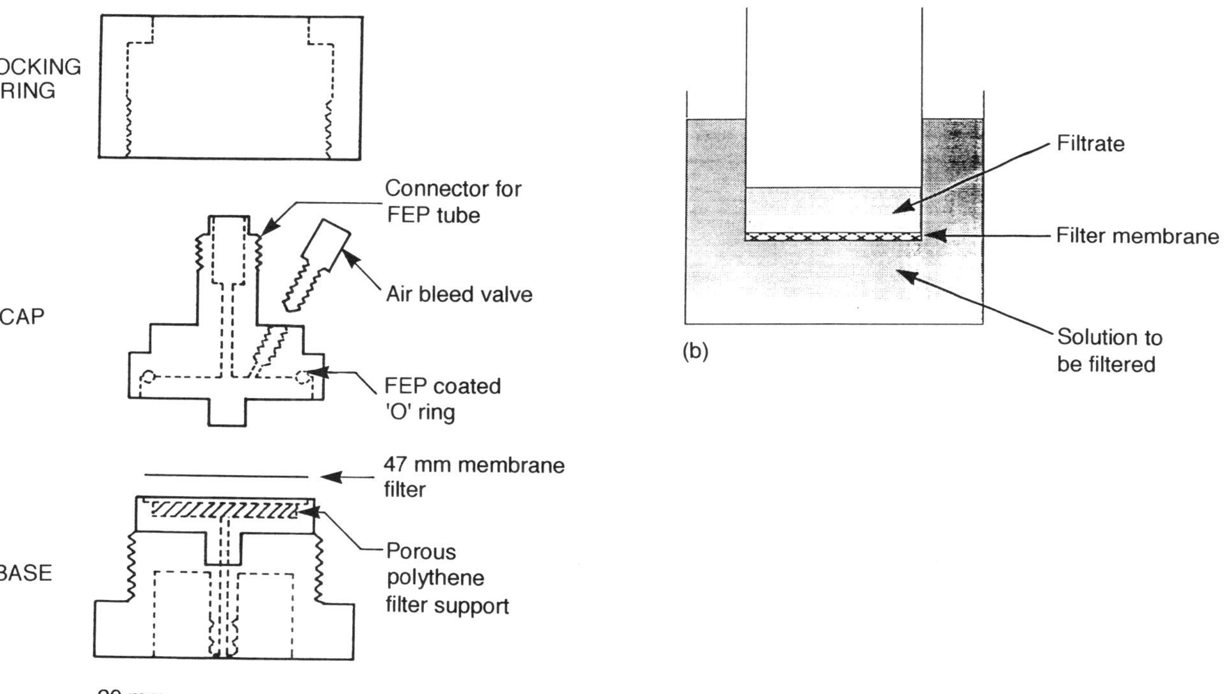

Figure 7.4 (a) An in-line filter holder for 47mm diameter membranes made entirely of PTFE and polyethylene and fitted with an FEP encapsulated 'O' ring (Morley et al.,1988; (b) filtration system for fragile organisms; the hydrostatic head gently forces the water through the membrane

A problem with both vacuum and pressure methods is that the cell walls of fragile biota can rupture during filtration, releasing their contents into the 'dissolved' phase. It is possible partly to circumvent this problem by limiting the pressure by using a gravity filtration system [Figure 7.4(b)] in which a small head of water forces the liquid through the filter. Contamination potential is relatively high, however, and flow rates are restricted. The approaches above using low vacuum or gas pressure are alternatives.

7.4.3 Continuous Centrifugation

Where large aqueous samples are to be processed, such as when suspended particulate matter is required from natural waters, continuous centrifugation procedures are appropriate (see for example Schüssler and Kremling, 1993), as they can now isolate particles of approximately the same size as the filters which have conventionally been used for this purpose. However, such systems (Figure 7.5) normally use a range of metallic components, and particular care is needed to limit contamination of both the dissolved and the particulate phase.

7.5 THE PRETREATMENT OF AQUEOUS SAMPLES FOR TOTAL ELEMENT ANALYSIS

In some analyses the chemical and physical forms of an element may not be well suited to the chosen method of determination. In natural waters, for example, colloidal material and complexes may not be in a form that can be collected by chelating ion-exchange resins. If a total element analysis is required, these unavailable forms must be converted to reactive forms. Acidification of the solution may dissociate complexes and destabilize colloids, but in other cases some other means of converting the analyte to a suitable form may be needed. The addition of a chemical oxidant, such as persulphate, followed by heating can destroy some organics. Such processes, however, involve the addition of reagents with associated contamination potential. There are a number of procedures which are inherently cleaner.

(i) Ultra-violet (UV) oxidation

Ultra-violet irradiation of aqueous samples can generate free radicals in solution, which can attack organic material and at least in part convert it to carbon dioxide. Samples are irradiated in tubes made of silica, which is inherently a non-contaminating material. Persulphate assists the photolysis of organic matter, but is at its most effective under neutral or alkaline conditions; this is incompatible with acidification, and metal losses through hydrolysis and adsorption on container walls can be high. Addition of the persulphate reagent may also add contamination.

(ii) Ozonolysis

Ozone in solution can be a powerful oxidant and can break down organic material.

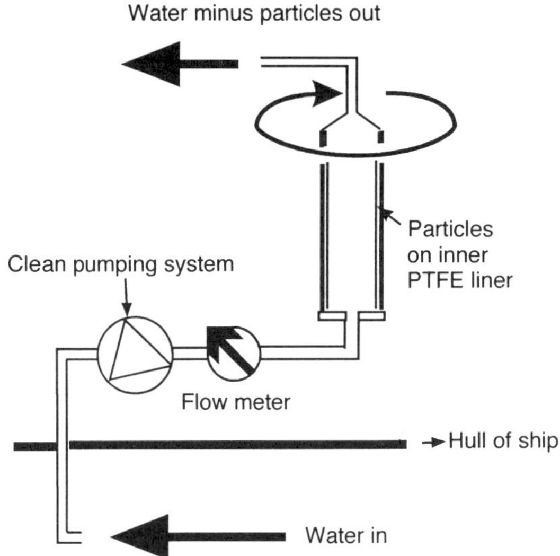

Figure 7.5 A continuous centrifuge system for the underway collection of marine particles (Schüssler and Kremling, 1993)

A stream of oxygen containing ozone from a generator is passed through the sample for an extended period of time. The same caution about the adsorption of analytes applies here as with UV techniques, and contamination problems may exist with the large volume of gas passing through the sample.

Neither method is likely to produce complete oxidation of organics in all samples, but the extent of decomposition may be sufficient to release the analytes of interest.

7.6 DRYING OF SAMPLES

Most types of sample contain at least some water. Drying may be required to permit accurate, reproducible weighing, or to prevent difficulties arising with water sensitivity in subsequent steps. Removal of water will also limit biological modifications of the sample during storage. Care must be taken to ensure reproducibility of drying as, in complex samples, the water loss may be a function of the conditions used. The release of water molecules locked in clay mineral lattices, for example, increases with the temperature used and the heating time. Two main drying techniques are available.

(1) *The conventional oven* The sample is put in an oven (typically close to 100 °C) and dried until constant weight is achieved. Drying can be aided by a fan, but the air stream drawn into the oven can also introduce airborne contaminants, dislodge any corrosion products, and cause contamination within the oven. Dried

samples should be kept in a desiccator to ensure that there is no take up of water from the atmosphere. The procedure is simple, but the dried sample can be a hard mass which is difficult to grind, and volatile analytes may be lost. A variant is the vacuum oven, where the drying action is enhanced by maintaining a low pressure in the oven cavity. The sample can thus be dried at a lower temperature, reducing loss of volatile analytes

(2) *Freeze drying* The sample is frozen and, when a vacuum is maintained, ice sublimes and the vapour is pumped away. The drying is done in a closed environment without a flow of air over the sample, which is therefore protected from oxidation and contamination. Any back-diffusion of oil vapour from the pump can normally be eliminated by a suitable trap. The equipment is more expensive than a conventional oven, but it produces a much more friable product and losses of volatiles and heat sensitive analytes are reduced.

7.7 THE GRINDING OF SOLID SAMPLES

The grinding of solid samples prior to analysis is done for a variety of reasons.

(1) A ground and finely divided sample allows effective homogenization and subsampling.

(2) Reducing the particle size increases surface area and facilitates attack during digestion procedures.

(3) For some analyses, such as X-ray fluorescence, a fine sample is needed to prepare a fused or pressed disc for analysis.

Although solid phase samples typically contain higher concentrations of analyte than dissolved samples, clearly there are many opportunities for contamination during grinding. The simplest apparatus is the mortar and pestle, and for trace analysis the most suitable type is normally made of agate. The mortar and pestle need careful cleaning between different samples, particularly when older equipment is used that has surface scratches which can harbour particles from earlier samples. A careful wipe with a damp non-particle-generating tissue is effective for most applications. In more demanding situations, after the initial clean, the grinding and then disposal of a small portion of the sample prior to the grinding of an analytical aliquot can reduce any carry-over. Contamination must be guarded against. There have been cases where jewellery rings worn by an analyst have introduced traces of metals into the sample, leading to highly erroneous geochemical interpretations. In mechanized mortar and pestle systems, particles from the motor and gear system must be carefully excluded from the sample. Conventional porcelain mortar and pestles can only be used in the least demanding analyses, where eroded material from the porcelain does not pose a contamination risk.

For brittle materials the diamond mortar has often been used. Here a steel rod slides in a close fitting metal sleeve with base; the sample is placed in the sleeve, and the rod is inserted and then repeatedly struck with a hammer. The action of the rod pulverizes the sample. Clearly this approach is of little use if the materials from which the device is constructed are the same as the analytes in the sample.

In the ball mill a ceramic pot contains the sample (which generally has to be relatively soft) and ceramic or tungsten balls. The pot is shaken vigorously in a mechanical device to disaggregate the sample. If tungsten balls are used the method is invalidated for this element, and potential contamination from the ceramic pot must be checked for.

In general, whichever technique is used, the more brittle the material, the more effective the break-up. Some materials can be made brittle by cooling to liquid nitrogen temperature. To ensure that all the sample has been effectively divided, it can be ground to pass through a sieve of known dimension; for sediment samples this may be a 63μm mesh sieve. Plastic meshes having a well defined aperture are needed; clearly metallic sieves are normally to be avoided. It is essential that all of the sample is ground to pass the mesh, as otherwise there will be fractionation of the sample and it will no longer be representative and homogeneous.

7.8 PREPARATION OF SOLID SAMPLES FOR DIRECT ANALYSIS

Techniques which allow the direct determination of analytes in solid samples frequently require special sample preparation procedures. For X-ray fluorescence (XRF) analyses the sample is often prepared as a fused or pressed disc of the powdered sample. Neither approach is totally without potential contamination problems. Fusion involves the addition of large quantities of additional reagents, and precautions must be taken to ensure that the dies used to form the compressed discs do not cross-contaminate the samples.

In neutron activation analysis, the solid samples must be presented to the thermal neutron flux in a container which does not contain any element that will cause problems with the subsequent analysis. This could occur through either contamination or the generation of activation products that would interfere with the gamma counting. As a consequence, the usual constructional material is either a polyolefin plastic or high purity silica. In specific analytical procedures, separations can be done either before or after the irradiation. Post-irradiation separations are to be preferred from the contamination viewpoint as only the activated contaminants are determined in the analysis. If manipulations are done before the irradiation, then the general strategies of contamination control and blank procedures must be adhered to.

7.9 DRY ASHING PROCEDURES

Such procedures are appropriate to samples with a high organic carbon content. The sample is heated and oxidized in a silica or porcelain crucible at typically 550–600 °C in a muffle furnace. The combustion residue is then normally leached with dilute acid or digested in strong acid to bring the analytes into solution.

The temperature will vary for each matrix. For food oils temperatures over 700 °C may be needed, but wheat products can be effectively ashed between 525 and 550 °C. The temperature is a compromise between maximizing the oxidation of the organic matter and minimizing the losses of volatile analytes. Whilst most elements, particulary those with involatile oxides, are retained at these temperatures, losses of S, Se, P, As, Sb, Ge, Tl and Hg are known. Additionally, elements forming volatile inorganic and organic species are prone to losses (e.g. Pb as $PbCl_2$, and organoarsenic compounds). 'Ashing aids' such as alkali nitrates can be added to aid decomposition, to reduce losses through the formation of non-volatile species, and to reduce interaction with container walls. All such additions, however, risk contamination from the added materials. In this type of open system it is difficult to control contamination, and this can be a major problem. Variable chromium and nickel blanks encountered in the analysis of blood, using a dry ashing procedure in a furnace, have led to an alternative wet ashing procedure being adopted.

7.9.1 Low Temperature Ashing

Oxygen molecules at low pressure (1–5 Torr) are excited by a radio-frequency field at 13.5 MHz to generate oxygen in the excited singlet form. The excited form has a lifetime of about one second, and is passed over the sample to be oxidized (Figure 7.6).

The sample must be close to the radio-frequency field because of the short lifetime of the singlet oxygen. The low temperatures involved, typically 150–200 °C, cause only low losses of volatile analytes, and the sample can be jacketed to limit the temperature even further. The method has been successfully used for As, Cd, Sb, Pb, B and Ge, but Hg is still lost. The method is inherently very clean, as the incoming oxygen can be filtered prior to use and no ashing aid is required. The technique is limited to relatively well dispersed samples, or requires periodic break-up of the sample, as only the surface of the material is attacked. If a PTFE container is used to hold the sample, the excited oxygen can attack the polymer and lead to the generation of atomic fluorine, which can aid the decomposition. This may lead to the loss of some elements by the formation of volatile fluorides. Losses are reported to be insignificant for Sn, Fe, Pb and Cr. The technique is slow, typically taking several hours per sample or even days for large samples.

7.10 FUSIONS

Fusions have long been used for the refractory materials (such as Ti and Zr ores, slags, and mixed oxides) which are difficult to attack by mineral acids. In a typical operation, the flux agent is added in excess to the finely divided sample (ratios vary from 2:1 to 50:1) held in a vessel which is able to resist the fusion conditions.

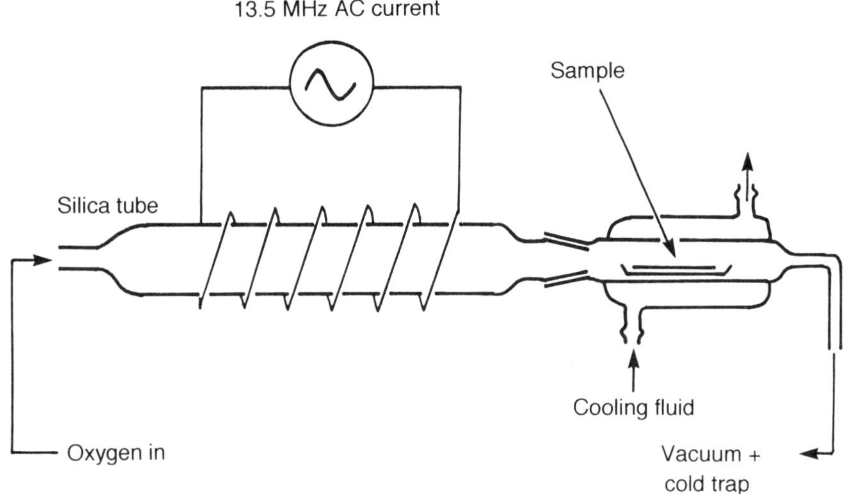

Figure 7.6 A low temperature ashing system

The vessel is heated in a muffle furnace to temperatures as high as 1200 °C. When the fusion is complete, the cooled solid is removed and dissolved in water or dilute acid to release the analytes into solution. A list of commonly used flux agents is given in Table 7.1.

Table 7.1 Commonly used flux reagents used in fusion techniques

Inorganic electrolyte	Notes
Sodium carbonate	Useful for silicates Can have added oxidizing agents (e.g. KNO_3) for elements such as Cr, As and Sb
Sodium hydroxide	Strong basic flux for e.g. silicates and silicon carbide N.B. Cannot use Pt crucibles
Sodium peroxide	On decomposition forms oxidizing flux; application to alloys of Ni, Cr, Fe, Mo and W, and to sulphides N.B. Cannot use Pt crucibles
Potassium hydrogen sulphate	On heating potassium sulphate and sulphur trioxide are formed. Useful for refractory oxides such as ZrO_2, and MoO_3.
Boric oxide	Appropriate when an acidic flux is needed in the determination of alkali metals

The fusion techniques are successful because the molten inorganic electrolytes are very powerful solvents in their own right. Reactivities and solubilities are increased at these higher temperatures, and the molten salts can act as Lewis bases and acids. At the high temperatures used, fusion techniques are also very effective at decomposing organic matrices.

Whilst fluxes are effective with certain types of samples, there are also major drawbacks. Because high flux to sample ratios have to be employed, the flux reagent can add significant amounts of analyte to the sample. There is no simple answer, as the materials which are typically used are difficult to purify. The vessels used are normally made of platinum or sometimes of nickel or silver. Although they are relatively inert, some dissolution of these materials can occur and lead to contamination; losses due to interaction of the analyte with the vessel wall can also occur. The flux reagent adds significantly to the dissolved salt content of the solutions which are generated, and thus may pose problems with subsequent analytical steps. Generally, therefore, these methods are to be avoided if possible when working with low analyte concentrations. Advances in plasma based techniques, in which suspensions can be introduced directly into the plasma, may offer one way in which the preparation of fluxes can be circumvented.

7.11 DISSOLUTION OF SOLID SAMPLES

Many analyses of solid materials require the sample to be brought into solution. This may be to allow a separation or other manipulation to be carried out in the dissolved phase or to provide a solution of sample for an instrument. Because of the manipulations and reagents which are involved, this is a stage in the analysis where losses or contamination can cause major changes in both the analyte concentration and the form in which it is present. There are only a few classes of materials, such as simple salts, where the sample will go directly into solution.

7.11.1 Use of Bases and Enzymes

Some biological tissues can be solubilized by quaternary ammonium hydroxides. Tetramethylammonium hydroxide (TMAH) in alcoholic solution is effective at bringing into solution a range of human and other tissues; moderate heating speeds the dissolution. The high pH can result in analyte losses on standing, and commercial preparations are available which contain a complexing agent to keep the elements in solution. Sample solutions can be introduced directly into flame and flameless atomic spectroscopy instruments, but great care is required with calibration. This is usually done by standard additions or using biological standard materials, because of the complex matrix.

Enzyme systems can also be used to bring tissue samples into solution but, as with the TMAH approach, the organic material is largely converted to simpler soluble organic substances and is not completely destroyed.

These are infrequently used techniques, and analyses yielding simple inorganic matrices have been preferred.

7.11.2 Partial Dissolution, or Leaching

In some cases it is informative to know the fraction of an element that is released from the sample under well defined conditions. From the point of view of potential toxicity, the amount of lead leached from a glazed plate by a dilute acetic acid solution is more informative than the total lead content of both the pottery and the glaze. In the study of marine and freshwater sediments, leaching procedures have been developed for a range of trace elements, in which a series of increasingly more vigorous chemical reagents sequentially release different fractions of the element. These studies are designed to give information on the geochemical associations and environmental availability of the elements in the solid. An example of this approach is shown in Table 7.2.

Table 7.2 The sediment leaching scheme of Tessier *et al.* (1979)

Components of sediment from which trace elements are released into solution	Reagent used
1. Exchangeable	Magnesium chloride; mass action effect
2. Carbonate	Acetate buffer at pH 5; carbonate dissolution
3. Iron and manganese oxide	Hydroxylamine hydrochloride; reduction of less crystalline oxides
4. Organic matter and sulphides	Acidic peroxide; attack of organic and sulphide phases
5. Residual	Nitric/perchloric/hydrofluoric total digest

Each step has its own blank problem, and the complexity of the experiment is compounded by the sequential approach. Data are rarely unambiguous in environmental interpretation, and the accuracy of each leach step is difficult to assess (Martin *et al.*, 1987).

7.11.3 Sample Dissolution using Acids

Some materials, such as carbonate rocks and some sulphides, can be brought into solution by simple dilute acids. Such samples, however, are comparatively rare and more rigorous methods normally have to be applied.

Sample dissolution by concentrated acids results from a number of factors. These include the high proton concentration, the oxidizing action of the acid (if any), and in some instances specific reactions or strong complexation with the acid anion. The term wet ashing is often used for these methods. Several acids can be prepared in sufficiently high purity for trace analysis (see Chapter 5). Each acid has particular features which can be used to advantage in a decomposition scheme.

7.11.3.1 Hydrochloric acid

On boiling the concentrated acid, which has an initial concentration of about 12 mol/dm^3, HCl is lost until an azeotrope mixture is formed with a concentration of ca 6 mol/dm^3. Hydrochloric acid is a strong acid, but it is not oxidizing. However, the high chloride ion concentration assists the dissolution of a number of elements, in particular Au, Tl and Hg, and to a lesser extent Fe, Ga, In and Sn, through complex formation. Hydrochloric acid can be used to dissolve carbonates, some sulphides and electropositive metals, and to extract some metals from silicate lattices (although the lattice itself will not be fully destroyed). Whilst most metal chlorides are soluble, the Ag and Tl salts are not, and Pb chloride is only sparingly soluble in cold aqueous solutions. Very high purity hydrochloric acid solutions can be made or purchased.

7.11.3.2 Nitric acid

Concentrated nitric acid is about 14 mol/dm^3 and contains between 65 and 69% of HNO_3; when higher concentrations of HNO_3 are present in the reagent it is called fuming nitric acid. Concentrated nitric acid is both a strong acid and a strong oxidizing agent.

> **Exceptional Risk!** Addition of nitric acid to some readily oxidized organic matter, and in particular solvents, can lead to an unpredictable and often delayed explosion or ignition

Dissolution of samples containing significant amounts of metallic Al, Cr, Ti, Nb and Ta is not possible because of the formation of insoluble metal oxide films which inhibit further reaction. Although the oxidizing action of the acid will effectively attack organic matter, complete destruction is rare, and so the acid is often used in combination with other acids. Unlike chloride ions, nitrate is only weakly complexing, and metals such as Sn, W and Sb can be hydrolysed and precipitate as hydrated oxides. Nitrate salts are soluble and, in combination with its other properties, this acid therefore offers the widest range of applicability.

Aqua regia

Aqua regia is a 1:3 (v/v) HNO_3:HCl mixture of the concentrated acids, which generates in solution very reactive oxidation products including chlorine and nitrosyl chloride. Classically the mixture is associated with the dissolution of noble metals, but it is also effective for biological materials. There may be some advantages in adding the nitric acid first to carry out a preliminary oxidation and then adding the hydrochloric acid.

7.11.3.3 Perchloric acid

When hot and concentrated, perchloric acid is a very powerful oxidizing agent. The acid forms an azeotrope containing 72% $HClO_4$ which boils at 203 °C, although the acid is more commonly handled in the safer 60% concentration.

Exceptional hazard warning!
Risk of explosion

It is very important to be fully aware of the explosion hazard associated with the use of this acid. Many inorganic and organic perchlorates are metastable and can explode when provided with sufficient energy by overheating, striking or even the breaking up of encrusted material with a spatula. Safety aspects associated with this acid, including the use of fume cupboards with washed down walls and scrubbed exhaust systems, have been discussed earlier in this book. The acid must be used with care and respect. If no alternative method is suitable, the acid should always be used in combination with other acids. This can serve two purposes, pre-oxidation and providing a high boiling point solvent. Easily oxidizable material should be first oxidized with another acid. Samples are normally heated with nitric acid alone initially to oxidize more tractable material before the addition of perchloric acid. Sometimes these acids are added together (e.g. 3 to 5 parts nitric to 1 part perchloric), as the perchloric acid does not become oxidizing until concentrated, at which time the nitric acid has been evaporated and the easily oxidized material removed. One of the most hazardous situations arises when a perchloric acid digestion goes partially dry, overheats and explodes. To minimize this risk sulphuric acid is frequently added, as it has a very high boiling point and will remain in the flask until all the perchloric acid has vaporized. When a sample is being heated with perchloric acid, it is important to ensure that significant quantities of dried material do not build up on the inside of the flask as the liquid evaporates. Such material should be washed back into the bulk of the liquid to minimize the chances of explosion.

7.11.3.4 Sulphuric acid

Concentrated sulphuric acid is about 98% (w/w) and has a concentration of 18 mol/dm^3. When hot and concentrated it is a strong acid and oxidizing agent; it is also a dehydrating agent, which can be useful in helping to destroy organic matter.

Carbon residues can be left if the acid is used on its own, and sulphuric acid is therefore normally used in combination with an oxidizing acid such as nitric acid, which leads to a cleaner digest. A major advantage of sulphuric acid is its high boiling point (330 °C). As sulphates are generally involatile, evaporation to dryness allows volatile components of the digest mixture to be driven off without losses of analyte. Thus excess HF or HCl can be readily removed. The sulphates of Pb, Sr and Ba are insoluble and Ca sulphate is only sparingly soluble, and so the analysis of these elements should not include sulphuric acid digestion steps.

7.11.3.5 Hydrofluoric acid

Although HF is non-oxidizing and a weak acid, it has the important property of attacking important refractory materials. The fluoride ion can form strong complexes with a range of elements, including those elements which form stable refractory oxides. The acid forms an azeotrope containing 36% of HF, which boils at 110 °C, and is available in high purity forms. The main application of the acid is to bring silicon compounds such as silicates into solution. The Si is lost in open digestions as the volatile SiF_4, and boron is removed as BF_3. Hydrofluoric acid is very rarely used in isolation; it is normally combined with acids which will complement its action on silicate materials. If organic components are present, nitric and perchloric acids could be used.

Exceptional hazard warning!
Severe burn risk

There are important practical points regarding the use of HF. It can cause very severe and painful burns, and can readily penetrate the skin because of its polar nature and similarity in behaviour to water. Full protective clothing must always be worn, and the integrity of gloves in particular should be checked before use. Hydrofluoric acid burn cream should always be on hand, and personnel using the acid, laboratory colleagues and first-aid personnel must be fully aware of actions to be taken in an emergency. Clearly glass and silica ware cannot be used in combination with this acid, and PTFE equipment is generally the material of choice.

General properties of the commonly used acids are given in Table 7.3.

7.11.3.6 Use of acid mixtures

Throughout the discussion of simple acid digests, examples have arisen in which a single acid does not have properties which will satisfactorily digest a sample. Most wet digestion procedures use more than one acid in combination. The most important characteristics and limitations of individual acids are given in Table 7.4.

Table 7.3 Properties of mineral acids commonly used in digestion procedures; data are for concentrations which are typically available

Acid	Boiling point (°C)	Weight %	Specific gravity	Concentration (mol/dm3)	Volume (ml) to be diluted to 1 litre to give an approximately 1 mol/dm^3 solution
Hydrochloric	109 (6 mol/dm^3 azeotrope)	35	1.18	11.3	89
Nitric	121 (67% HNO_3 azeotrope)	70	1.42	16	63
Perchloric	203	70	1.66	11.6	86
Sulphuric	330	96	1.84	18	56
Hydrofluoric	111 (36% HF azeotrope)	46	1.15	26.5	38

Table 7.4 Important characteristics and limitations to be considered in the choice of acid mixtures for digestion procedures

Acid	Characteristics	Limitations
Nitric	oxidant, forms soluble salts	volatility
Hydrochloric	forms strong chloride complexes	volatility
Sulphuric	oxidant, high boiling point	residual carbon with organics
Perchloric	strong oxidant	exceptional explosion hazard
Hydrofluoric	removal of Si and B	exceptional burn hazard

The complete digestion of a soil sample for trace metal analysis might therefore involve hydrofluoric acid to remove the silicon, perchloric acid as an oxidant of organic matter and sulphuric acid to ensure that the perchlorate salts do not present an explosion hazard.

It should be remembered that the objective is always to bring the analyte quantitatively into solution; this does not always require complete dissolution of all components in a sample. Examples of acid mixtures appropriate to a range of materials including plant and animal tissues, geological materials and environmental samples are given in Table 7.5.

7.11.3.7 Equipment used for dissolution procedures

When rigorous digestion procedures are being used, it is essential to have equipment made from materials which will not contaminate the sample or interfere with later steps in the analysis due to, for example, bonding of sample residue to the container surface. Also the material must be stable at high temperatures. In less rigorous applications borosilicate glass may be suitable, but the most commonly used materials are high purity silica and PTFE or other similar fluorinated polymer. Containers used for high analyte content samples should not generally be used subsequently for samples having very low analyte concentrations. It should be particularly noted that the white sintered PTFE used to make beakers is relatively porous, which can lead to the retention of small amounts of analyte, even after apparently rigorous cleaning.

7.11.3.8 Use of closed decomposition vessels

In the past much work has been done in fume cupboards with open digestion vessels. The sample is weighed into the clean container, the digestion mixture is added and the container is heated on a suitable hotplate; the procedure may include a reflux stage in which the container is covered and the cap acts as an air condenser. This type of procedure has much potential for contamination. The air flow through the fume cupboard is continually bringing dust from the general laboratory which may become entrained in the digestion vessel. Additional sources of contamination can include corroded metal hotplates and fume cupboard compo-

Table 7.5 Examples of different mixtures of acids used for the digestion of a variety of samples (taken from Van Loon, 1985)

Sample	Acid mixture used for digestion	Elements determined	Determination method	Notes
Soil	Nitric Perchloric Hydrofluoric	Ca, Al, Ni, Mn, Zn, Co, Cu, Cr, P	Inductively coupled plasma atomic emission spectrometry (ICP-AES)	Part of the digestion is done under pressure.The amount of HF can be increased if the sample contains a high proportion of silica, such as with high sand soils
Botanical samples	Nitric	Mo (GF-AAS)	Graphite furnace atomic absorption spectrometry need for additional acid	Good recovery from certified reference material indicated no treatment
Sewage sludge	Nitric Hydrochloric (aqua regia)	Cd, Cr, Cu, Fe, Mn, Ni, Pb, V, Zn	AAS	Generally >90% of the total metals were released by this attack. This recovery was considered adequate for the use of the data
Blood	Nitric Sulphuric	Se	Hydride generation AAS	
Fish and other tissue, plant material and soils	Nitric Sulphuric	Hg	Cold vapour AAS	The digestion solution must be kept oxidizing to prevent losses of Hg^0. Vanadium pentoxide is added to aid the oxidation

nents and inorganic particles from sand baths. Chapter 2, The Working Environment, describes cleaner environments.

An alternative strategy is to use closed decomposition vessels, which are often termed 'bombs'. These isolate the digestion process in an enclosed container that can withstand the pressures developed during digestion, and effectively exclude external contamination. An additional major advantage is that the increased pressure enhances the rate of decomposition.

A typical bomb consists of a metal cylindrical cup and screw cap incorporating a pressure release valve, and a PTFE liner and top (see Figure 7.7). The sample and acid(s) are placed in the liner, the unit is closed, and the unit is then heated either in an oven or a water bath for some tens of minutes. After cooling, the contents are removed. Caution is required to ensure that there is no contamination from the metal jacket. At elevated temperatures and pressures diffusion of acids through PTFE is enhanced, and this can lead to corrosion of metallic components.

Some designs do not have a metal jacket but have very thick PTFE walls to help reduce distortion of the container and consequent leakage; others use acrylic jackets. These all-plastic designs are restricted to lower working temperatures. The safe working sample size, which is dependent on matrix and on the volume and type of acid used, is specified by the manufacturer.

> Do not exceed recommended sample size and acid volumes for bombs

7.11.3.9 Microwave digestion systems

Microwave heating of the contents of closed digestion vessels is rapidly being adopted. The approach has many advantages. The digestion is quicker, as microwaves heat the sample directly. In a conventional bomb the external jacket has to be heated first, and the sample/acid mixture then warms by conduction. Microwaves can also aid the decomposition by breaking up the sample surface. Depending on the size and type of sample, the digestion can be very quick, and typically takes only a matter of minutes for small samples.

The digestion containers must be transparent to microwave radiation, and should not contain metal components. They are typically made from PTFE or PFA materials, and have built into the design a pressure release or rupture valve. In some systems a central collection pot is provided to collect any digestion mixture that escapes. Some sophisticated systems have a sensor on a digestion vessel which gives information on pressure that can be used to control the power level in the oven. The use of inherently clean materials and avoidance of metallic components in bomb construction are advantageous in limiting contamination potential. However, such digestion vessels have a limited lifetime owing to changes in shape

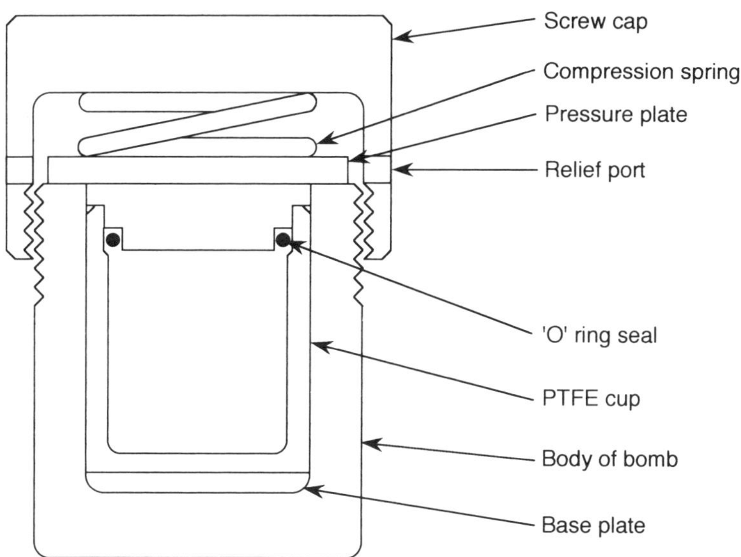

Figure 7.7 A bomb for the decomposition of samples at elevated temperatures and pressure

and stress developing over repeated heating cycles. Commercial systems are available, although several laboratories use domestic systems that have been modified to ensure adequate ventilation of the main chamber and to remove possible sources of contamination. The use of microwave digestion systems is discussed in detail in Kingston and Jassie (1988).

7.12 SEPARATION AND PRECONCENTRATION

Under ideal conditions the analysis of trace elements would not require any initial concentration or separation stages. However, even with recent developments in selectivity and sensitivity, direct instrumental analysis is often not possible because the concentrations of analytes are below the limits of determination of the method or the matrix provides an interference which must be removed. Both problems may be encountered simultaneously in a sample. Separation and concentration procedures are therefore frequently components of an overall analytical scheme, either separately or in combination. These manipulations must provide a final sample that is optimized for the determination step.

Concentration and separation are integral parts of the determination step in the important electrochemical techniques anodic stripping voltammetry (ASV) and cathodic stripping voltammetry (CSV). In the first stage of ASV determinations, a high negative potential below the hydrogen over-potential is applied to a mercury or mercury coated electrode, and metal ions with a suitable reduction potential are collected from solution as amalgams with the mercury. In CSV, element complexes are formed in solution and then allowed to physically adsorb onto a mercury hanging drop electrode. After a set deposition time, they are then reduced by a cathodic increasing potential, and the current signal can be related to analyte concentration.

All additional steps involving separation and concentration in an analysis introduce the potential for contamination and losses, and the actions which can be taken to minimize such problems are discussed in this chapter. The methods used to separate quantitatively the elements in an analysis also often parallel important techniques for the purification of reagents (see Chapter 5).

Normally the separation of an analyte from its matrix or other analytes involves a transfer between two different phases. This two phase separation concept can give a useful categorization of the major separation techniques, and provides a basis for the discussion here (Table 7.6).

Table 7.6 Phase separation procedures used in analytical sequences for trace inorganic elements, with examples (after Miller, 1975)

Phase 1	Phase 2		
(contains or is sample)	Gas	Liquid	Solid
Gas	Diffusion (NH_3)	GC[a] (separation of metal complexes	Amalgam formation (Hg collection on Au)
Liquid	Distillation	Solvent extraction (metals in solution) Ultrafiltration	Ion-exchange Electrode deposition (ASV[b]) Co-crystallization Co-precipitation
Solid	Sublimation	Selective leaching (metals in sediments)	

[a] Gas chromatography.
[b] Anodic stripping voltammetry.

The focus of this chapter is on methods which are currently used for separation and concentration, their limitations and advantages, and how a thinking approach and an appreciation of the principles behind good trace analysis can improve their application. Separation science is a major field in its own right, and no attempt is made here to cover this subject rigorously; background theory can be found in the book by Miller (1975), whilst reviews of separation and concentration methods are given by Bachmann (1981), and Mizuike (1983). There are continual developments in separation and concentration techniques, including their direct interfacing with instrumentation. Despite such changes, most of the basic principles discussed in this chapter will still be appropriate in the future.

7.12.1 Evaporation

This approach is appropriate to aqueous or organic samples where the total solute concentrations are low and the concentration process does not give an unacceptable matrix for the element to be determined. The technique is suitable for snow or freshwater samples and for high purity reagents undergoing checks on elemental content, but cannot be applied to samples such as seawater, where the high salt content leads to major matrix problems in the subsequent instrumental analysis. The prime requirement is for an evaporation environment which is free from atmospheric contamination. The use of either a closed filtered air container or a clean wet station, as discussed in an earlier chapter, is essential for samples with a low concentration of analyte. As is normally the case, rigorous blanks must be run to check for general contamination and for problems from individually contaminated containers.

7.12.2 Volatilization

Where the sample is more volatile than the matrix, or can be converted to be so, this property can be used to good effect in a separation procedure.

Volatilization is an elegant technique for the preconcentration and separation of mercury from a broad range of materials, from digested rock to tissue and aqueous samples. The ionic mercury is reduced to the metallic state, in which form it is readily volatilized from aqueous solutions. Once purged from solution in an inert gas stream, the mercury can either be measured directly by cold vapour AAS or atomic flourescence spectrometry (AFS) or pre-concentrated as an amalgam with gold. Mercury is quantitatively removed from the gold by rapid and reproducible heating, and the resulting pulse of mercury is determined as above. The gold can be present as fine wire, mesh, or as a coating on silica or other inert substrate. Where the mercury is already in the gaseous phase, as in the atmosphere, only the preconcentration step is required. Detection of organic mercury species requires different collection and separation procedures prior to gas chromatographic separation and detection. Control of contamination and losses is of paramount importance when dealing with this element. The problem is increased with organic forms, where concentrations are typically very low and several manipulations may be necessary in an analysis. The analysis should be designed to minimize these steps! The volatility of metallic mercury leads to ingress and losses of the element with containers made of mercury-permeable materials. Conventional polyolefin polymers have not been used, and containers are most frequently made of FEP and borosilicate glass. Losses of mercury metal can be prevented by oxidation to the ionic non-volatile Hg^{2+} form. A review of the determination of mercury in environmental samples, emphasizing the necessity for stringent contamination control to obtain accurate data, is given by Drabaek and Iverfeldt (1992).

The elements As, Sb, Se, Te, Sn, Ge and Pb form volatile covalent hydrides, and this property has formed the basis of efficient separation and preconcentration techniques for these elements. The element in solution is treated with a powerful reductant, usually sodium borohydride, and the resulting hydrides are swept from solution in an inert gas stream. If the concentration is high enough, the analyte can be determined directly by AAS or by ICP emission techniques. The analyte has to be in solution in a form which can react with the borohydride. Pretreatment of the sample to release complexed or otherwise bound forms of the element is therefore necessary if a total analysis is required.

Contamination potential is reduced when handling steps are reduced. The use of a single step, single container separation is thus desirable, and is demonstrated by the separation and concentration of elements such as Se, Te, Bi or Tl from non-volatile matrices by selective volatilization (Tschöpel, 1992). The sample is mixed with sodium chloride or vanadium pentoxide and introduced into the bottom of a high purity silica tube, which is placed in a furnace at about 1000 °C (Figure 7.8). The more volatile elements are driven off as chlorides or oxides and condense in the cooler part of the tube; they can be recovered in microlitre volumes of hydrochloric acid.

7.12.3 Co-precipitation

Elements are rarely present in trace analysis at concentrations which are sufficient to permit them to be directly and quantitatively precipitated from solution. Under suitable conditions, however, precipitates of other materials can quantitatively remove trace inorganic constituents from the aqueous phase. There are three principal mechanisms in operation in this co-precipitation process:

surface adsorption: the surface charge on the precipitate can attract ions in solution of the opposite charge;

inclusion: the analyte may isomorphically replace an ion in the crystal structure of the precipitate (mixed crystal), or be incorporated non-isomorphically (solid solution);

occlusion: ions are physically engulfed in the forming precipitate before they can diffuse or be carried away. This mechanism tends to be much less efficient than the other processes.

Co-precipitation can be highly efficient, and a wide range of analytes can be collected, but there are major disadvantages. The precipitate, which has a mass many orders of magnitude greater than that of the analyte, can be a major source of contamination. The removal of the analyte from the precipitate matrix may require further separation, with associated losses and blanks. There is also the contamination and loss potential in all the handling steps, such as the filtration and the dissolution of the precipitate.

Some of these problems can be negated by the use of organic reagents to complex and then to precipitate the analytes. The organic reagent is normally added in a water-soluble volatile solvent to an aqueous sample, and if precipitation of the

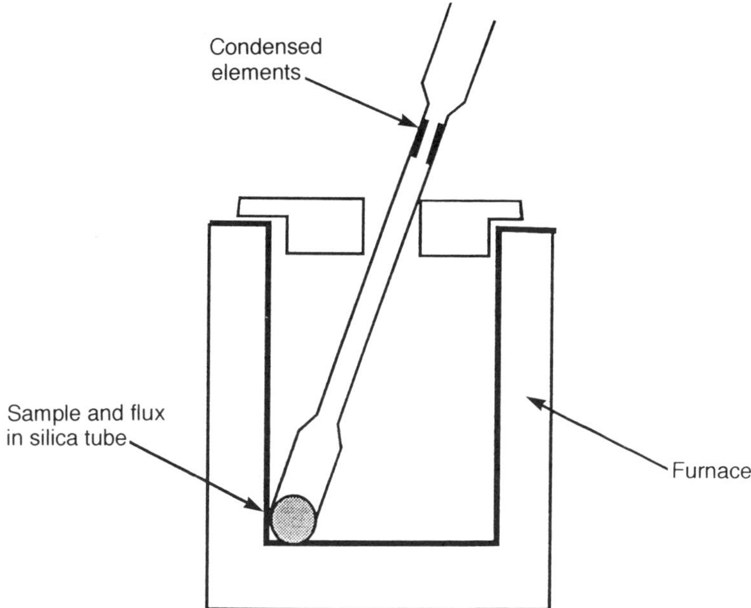

Figure 7.8 High temperature separation of elements using a silica capillary tube and furnace (Tschöpel, 1992)

organic complexant does not occur spontaneously, the organic solvent can be removed by heating to initiate precipitation. The organic precipitate can be destroyed by combustion or by acid digestion after collection by filtration. The removal of a range of metal complexes from natural waters by co-precipitation with a cobalt pyrrolidinedithiocarbamate complex, for example, is a further variant on this theme.

Despite the limitations, inorganic precipitates have been used in many applications. Iron co-precipitation can be employed, for example, for the concentration of rare earth elements from seawater at ng/l concentrations. A high purity iron solution can be prepared by the dissolution of spectroscopic grade iron in high purity hydrochloric acid, the hydroxide then being formed by treatment with sub-boiling distilled ammonia solution.

Trace anions as well as cations can be removed from aqueous solution. Phosphate, for example, can be collected from seawater by increasing the pH to precipitate $Mg(OH)_2$. This carries down a number of phosphorus species.

7.12.4 Solvent Extraction of Element Complexes

Solvent extraction is a very important technique for the preconcentration and separation of trace elements from a wide range of matrices. A complex is formed with the element of interest, normally in an aqueous solution, which will partition into a mutually insoluble or sparingly soluble solvent. The volumes by Stary (1964) and Morrison and Freiser (1957) contain much fundamental and useful information. Cresser (1978) provides a useful overview of the relevant theory.

The distribution of a species X between aqueous (aq) and organic (o) phases, is described by the partition coefficient P:

$$P = \frac{\gamma_{x_o}[X]_o}{\gamma_{x_{aq}}[X]_{aq}}$$

Gammas with subscripts, and square brackets, respectively denote activity coefficients and concentrations for the analyte species in each of the solvents.

This relationship only holds rigorously for well defined thermodynamic systems. In real systems, many species of an element, such as a range of inorganic and organic complexes, may be present simultaneously. From an analytical viewpoint the analyst is interested in the total amount of element that is transferred from one phase to another, and so the more empirical distribution ratio D is normally used, where:

$$D = \frac{[X_{total}]_o}{[X_{total}]_{aq}}$$

The X_{total} term includes all species of the element present in the phase considered. The distribution ratio can be measured readily, even for complex systems, and can be used in the optimization of a procedure. The amount of analyte transferred from the aqueous to the organic phase, expressed as the extraction efficiency E, is given by:

$$\%\text{Extraction} = \frac{100D}{D + \left[\frac{V_W}{V_O}\right]} = E_o$$

where V_w = volume of aqueous phase and V_o = volume of organic phase.

A very important aspect of this equation is that the extraction efficiency is independent of analyte concentration. This feature of the relationship has been tested down to picomolar concentrations of lead in seawater. The equation also shows that high extraction efficiencies are favoured by low V_{aq} to V_o ratios. There is, however, a practical limit to this ratio if a high concentration factor is required, and in trace analysis the addition of large volumes of solvent may have an associated blank. The use of sequential extraction steps can reduce the overall volume of solvent used, yet still maintain an overall high extraction efficiency.

7.12.4.1 Types of compounds separated by solvent extraction

The first step of the analysis is to generate, if not already present, a non-polar

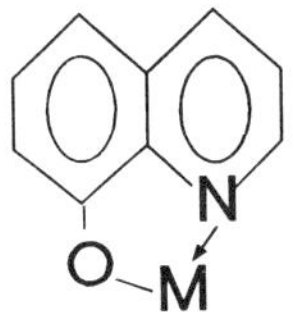

8-Hydroxyquinoline
(oxine; 8-quinolinol)

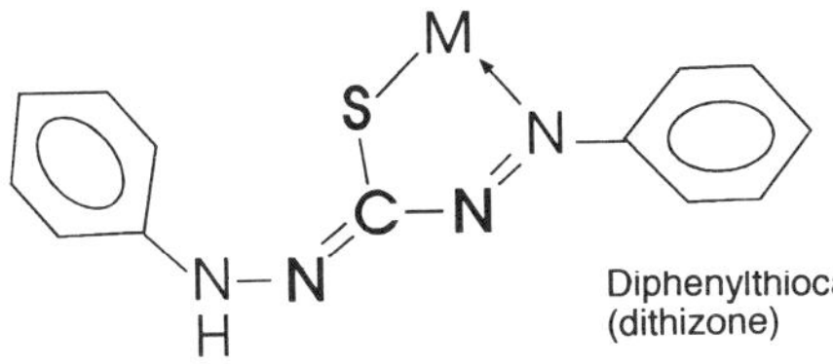

Diphenylthiocarbazone
(dithizone)

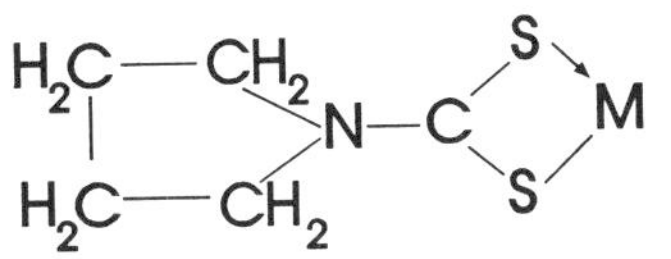

Ammonium pyrrolidine
dithiocarbamate (APDC)
(ammonium ions displaced
by metal in figure)

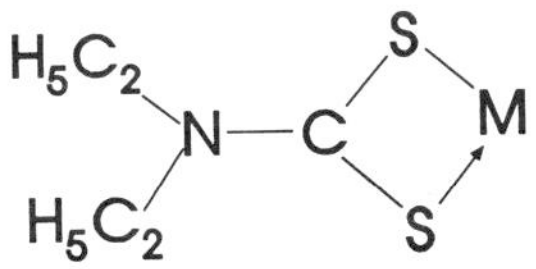

Diethylammonium
diethyldithiocarbamate
(diethylammonium group
displaced by metal
in figure)

Figure 7.9 Examples of structures of metal ion chelate complexes used in solvent extraction procedures for trace elements

species which will effectively partition into the immiscible organic solvent to be used. The main categories of solvent extractable species are as follows.

Naturally occurring complexes

Elements can already exist in association with naturally occurring organic complexants such as those found in a natural water.

Inorganic complexes

Complexes can be formed with simple inorganic ligands. Iron(III) will complex with chloride ions in strong HCl solutions (typically > 5 mol/dm^3) to form $Fe(Cl_4)^-$. This species can then be extracted into diethyl ether as an ion-pair

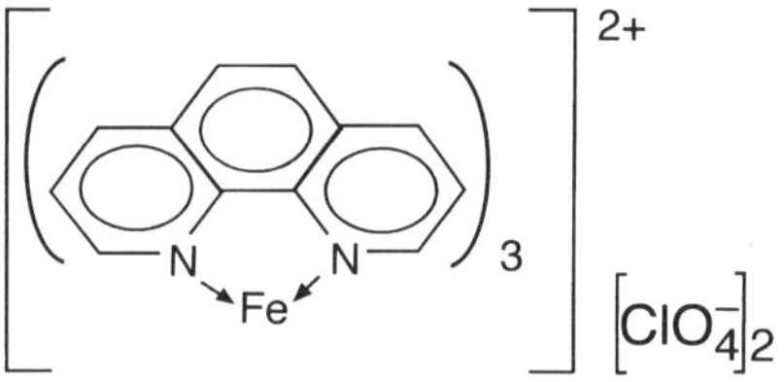

$$\left[\left(\text{phen}\right)_3\text{Fe}\right]^{2+}\left[ClO_4^-\right]_2$$

Figure 7.10 The solvent extractable ion-pair of the 1,10-phenanthroline–Fe(II) complex and perchlorate ions

Colloidal heteropoly acids can also be solvent extracted. The phosphomolybdic acid formed in the colorimetric analysis of soluble reactive phosphorus is normally reduced to phosphomolybdenum blue, a colloidal material that can be solvent extracted into 2-methylpropan-1-ol.

Multidentate chelate complexes

This is by far the most important route for obtaining complexes that can be solvent extracted from aqueous solution. The element chelates formed must be apolar to be efficiently transferred to the organic phase. If the chelating molecules fully balance the charge on the ion then transfer to the organic phase should occur. The coordination arrangements of several of the most commonly used complexing reagents are shown in Figure 7.9.

If a residual charge exists, the formation of an ion-association complex can sometimes be used to produce the required non-polar entity. An example is the ion-pairing between 1,10-phenanthroline–Fe(II) and perchlorate ions shown in Figure 7.10.

> The complexing agent as taken from the reagent bottle must always be assumed to be contaminated unless demonstrated otherwise

If particulate material is present in the complexant solution, this must be removed by filtration prior to further purification.

7.12.4.2 Control of pH

Control of pH is frequently necessary to ensure that conditions are favourable for the formation of a complex and that the resulting complex is in a suitable chemical form to partition into the desired phase. This is particularly important when a wide range of elements are to be separated at the same time. Buffering ensures that there is no major fluctuation of pH during the sample manipulations. Occasionally a matrix, such as seawater, may have sufficient inherent buffering capacity.

Buffers can be a significant source of contamination

Great care is required in trace analysis that no contamination is introduced with the buffer. The smaller the amount of buffer added, the lower the contamination potential but a reduced buffer capacity will be available. A range of buffer systems has been used, and techniques for reducing contamination to acceptable levels are discussed in Chapter 5, Reagents. The most important procedures for reducing contamination from buffers are to:

(1) remove any particulate contamination by filtration;

(2) use high purity reagents, such as a mixture of sub-boiling distilled acetic acid and isothermally distilled ammonia for the production of an acetate buffer;

(3) use an appropriate decontamination procedure; addition of the complexant to be used to the buffer and solvent extraction of metal impurities as their chelates is normally very effective;

(4) use rigorously cleaned storage vessels for the purified reagents.

Where samples are acidified for storage, or are the product of a digestion procedure, partial neutralization with base before addition of the buffer may be necessary. In order to assess the quantity of base which must be added, it is essential that checks should be performed on small sub-samples of the solution to be analysed. The volumes needed for the bulk of the sample can then be scaled up from the pilot experiment. Indicator papers and solutions, and electrodes used for pH measurement, are all potential sources of contamination, and should not be introduced into the sample to be analysed.

7.12.4.3 Other experimental variables

The rate at which material is extracted into the organic phase can be difficult, if not impossible, to predict, and this aspect is best determined experimentally. Reproducibility of mixing is improved when a mechanical mixing table is used. This has the additional advantages of freeing the analyst for other tasks and allowing more than one separation to be carried out at a time. Great care is needed to isolate the sample from any contamination associated with the mechanisms involved.

The choice of solvent is dependent on many factors. An ideal solvent should:

(1) have an appropriate density greater than that of water if the sample is to be drawn off, or less than that of water if the organic phase is to be aspirated into an instrument or otherwise directly accessed;

(2) have a very low solubility in water;

(3) be compatible with the next phase of the analysis;

(4) be able to dissolve high concentrations of the metal chelates of interest;

(5) form limited or no interfacial emulsions;

(6) be non-toxic and environmentally acceptable;

(7) be produced in an acceptably uncontaminated state.

In flame spectroscopy ketones and esters are most commonly used. Chlorinated solvents are normally avoided because of the health hazards associated with both the solvents themselves and the toxicity of flame combustion products (such as phosgene).

When additional purification of the analyte is required, or the sample has to be in aqueous solution for the determination stage, back extraction of the analyte can be used. Back extraction into a dilute aqueous acidic solution also ensures that most elements are stable over long periods of time, and thus prevents problems with any unstable metal chelates in the organic solvent. The advantages of the extra manipulations often outweigh the potential problems with an additional step in the analysis. Several procedures using back extraction have been applied to the determination of trace metals in natural waters at nanomolar and picomolar concentrations. The back extraction can be achieved either by addition of small aliquots of acid to the organic phase containing the chelate, or by evaporation of the solvent and oxidation of the organic residue by high purity nitric acid prior to dissolution in dilute acid for the determination step.

The pre-concentration step must be compatible with the determination step

High purity FEP is frequently used for the construction of funnels for solvent extraction procedures. The surface properties of these funnels change after extended use with dithiocarbamate reagents. Gradually the surface becomes opaque, and as this change progresses, the degree of carry-over of analyte from one extraction to the next increases. The resulting blank is most obvious when a low concentration sample is run directly after a high concentration sample. This partitioning of sample onto the walls is difficult to prevent, and simple solvent or acid rinses are not effective. Periodic replacement of the funnels is the most effective solution.

It is often assumed that a total measurement is made. For systems where a rigorous pretreatment has been undertaken, the analyte will be in a simple inorganic form amenable to chelation and extraction. However, in some samples there may be natural ligands which compete for the elements. Because of the very high stability constants of the commonly used chelates with metals, and the high concentrations of reagents used, the conventional wisdom is that a good measure of total metals is obtained. The competition of natural ligands for metals can be used to advantage in studies of natural complexing materials, such as those using cathodic stripping voltammetry.

A major limitation to solvent extraction procedures is the large number of steps and pieces of apparatus needed, with the associated contamination potential. The comfortable maximum volume to be extracted by hand shaking is about 1 litre. Continuous extraction systems have been developed, but these are often cumbersome to use. Even 1 litre samples may be inadequate if several analytes are to be determined individually, or if analyte concentrations are very low.

7.12.5 Solid Phase Collection Systems

The aqueous sample with analyte is allowed to interact with a stationary phase, and the analyte is then separated from the solid matrix. If removal of analyte from the stationary phase is subsequently carried out using only a small volume of reagent, concentration is also achieved. This group of techniques includes the use of chelating ion-exchange materials and the adsorption of metal chelates from solution. A recent review of this analytical approach and its application in the measurement of trace metals is provided by Kantipuly et al. (1990).

7.12.5.1 Adsorption methods

In this technique the metal chelate is formed in solution and then collected on an inert adsorbent, which is normally held in a column. The element can be eluted in an organic solvent as the complex, the solvent is then evaporated off, the complex is destroyed by acid digestion, and the element is taken up in dilute acid prior to determination. In other cases the element can be eluted directly in dilute acid, thus reducing handling steps. Complexes of metals with 8-hydroxyquinoline and dithiocarbamates have been collected from a seawater solution onto a material consisting of C-18 hydrocarbon chains immobilized on a silica substrate, and the macro-reticular resin Amberlite XAD-4, respectively. A problem with the silica material is instability at high pH, leading to its dissolution and the release of contaminants such as iron from the matrix.

7.12.5.2 Ion-exchange and chelating polymers

These are a range of materials having functional groups or chelating groups immobilized on the surface of an inorganic or organic support material. Their general properties and use have been discussed in the section on reagent purity in Chapter 5. Conventional anion- and cation-exchange resins have found only limited application in analytical trace metal analysis because of their lack of specificity. With carefully manipulated conditions they have, however, found uses such as the separation of rare earth elements from the iron used in their preconcentration and from barium, which interferes in their determination by mass spectrometry.

One of the earliest groups of chelating materials to be developed comprised the iminodiacetate modified polymers, such as Chelex-100. Their main advantage is a selectivity for polyvalent metals over commonly occurring monovalent cations such as Na^+ and K^+. As chelates are formed rather than simple ionic bonds, the overall rate of reaction is slowed. In analytical applications the resin is normally used in the ammonium or calcium form, rather than the hydrogen ion form originally suggested. If the hydrogen form is used, high salt content solutions displace protons from the resin, leading to a low pH solution passing through the column, which prevents the analyte in this aliquot of sample being retained. Once analytes are associated with the resin, and excess matrix ions have been rinsed from the column, the elements of interest are eluted from the column with dilute acid. A practical consideration is the change in size of the resin with counter ion (Table 7.7). Care must also be taken that 'channelling' or drying out of the column does not occur.

Table 7.7 Change in volume of Chelex-100 resin with different counter ions

Counter ion	Relative volume
H^+	0.34
Na^+	1.00
NH_4^+	1.25

Chelex-100 has found many applications in the collection of metals from natural waters. The optimum solution pH for the retention of a range of commonly determined trace metals from seawater is about 5.5.

It should not be assumed that such resins will collect all forms of a metal from solution. The iminodiacetate complexes are relatively weak compared with some of the organic—metal complexes which can be present in natural waters, and the prior destruction of complexes by ultra-violet irradiation or other treatment may be necessary. Additionally, the porous nature of the beads, inside which most of the chelating sites are located, means that colloids are excluded from the analysed fraction. The effective pore size of the resin (*ca.* 1.5 nm) must be compared with the less than 0.4 μm cut-off which is normally used for the definition of dissolved species in natural waters; colloidal forms will therefore not enter the beads. These properties of Chelex-100 may be used to advantage in analytical schemes which study the forms of metals.

Iminodiacetic acid groups are not the only complexing groups which have been attached to polymeric backbones. One of the more commonly used systems is 8-hydroxyquinoline bonded to silica or glass. This is more specific and forms stronger complexes with many trace metals. The flow rates of sample through the column can also be greater. A range of elements can be collected from natural aquatic samples, and the method has been applied to Cd, Co, Cu, Cr, Mn, Ni, Pb and Zn in seawater at pH 8. The stability of the substrate can be greatly improved by bonding the 8-hydroxyquinoline onto an organic polymer.

The development of immobilized complexing agents such as tetra-azamacrocycles and polyacrylamidoxime is an active area of development. Key criteria for the application of these and related materials in the separation of trace inorganics are:

(1) high selectivity for the analytes of interest;

(2) rapid flow rates with quantitative collection of analytes;

(3) ease of elution of analytes, preferably in small volumes of reagents in which the analytes are stable and ready for direct interfacing with the determination step;

(4) stability of the support resin over the range of pH conditions used in the analysis;

(5) robustness; ideally the resin material should be reusable;

(6) low and reproducible blanks to give good detection limits.

Chelating resins can be used in either column or batch mode. Columns have advantages because the resin material is well isolated. In batch mode the resin must be added to the sample, mixed and recovered, the analytes then being eluted; these further transfer and handling steps all have associated contamination potential. The materials used for column construction should be of high purity and carefully cleaned before use. Glass wool should not be used to support the column of resin, as the high surface area and inherent lack of cleanliness can lead to losses and blanks. Conventional glass frits can also give problems. The best material for supporting the resin is sintered polyethylene, which can be readily cut to size. The column is also best made of plastic, and commercial units with fitted plastic frits are readily available. To ensure good and reproducible behaviour of the resins, it is important to optimize and control carefully the flow of sample through the column. Crude control can be achieved by a tap on a reservoir attached to the top of the column, whilst much better regulation is achieved by using a peristaltic pump. This pump must be downstream of the column, as the vinyl and other plastics commonly used for pump tubing may be sources of contaminants. The downstream tubing can be removed after the sample has been passed, and normal washing of the resin and elution of trace elements can be carried out under gravity.

7.13 THE DETERMINATION STAGE

A description of the wide variety of determination methods that exist for trace analysis is outside the scope and objectives of this book. However, as the overall quality of an analysis is only as good as the weakest link in the chain of procedures involved, including the determination step and the presentation of the sample to the instrument, these last steps also need critical appraisal if acceptably accurate and precise data are to be obtained. In the future we will be moving towards the ideal where sampling, storage, preconcentration and laboratory determinations are minimized or eliminated by the use of highly sensitive and selective *in situ* analyses in, for example, the environmental, biological and industrial fields. However, for the immediate and intermediate future, we will continue to apply the more complex procedures commonly in use today.

Some analyses may need only very limited work-up prior to the determination step, whilst in others there are extensive processing steps to provide a sample with the analyte in the form and at a suitable concentration for the determination step of choice (e.g. chelation–solvent extraction prior to graphite furnace atomic absorption spectrometry). In a general sense, the less the preconcentration prior to the determination step, the lower the analyte concentration and the greater the potential for contamination during presentation of the sample to the instrument. With anodic stripping voltammetry (ASV) and cathodic stripping voltammetry (CSV) procedures, where preconcentration is done in the electrochemical cell, it is thus imperative that electrodes and cell assemblies be fabricated from non-contaminating materials and cleaned to the same extent as containers for the raw samples. Clearly the use of carefully designed blank experiments to test for contamination is very important here.

Even when a preconcentration step has been undertaken, there is potential to contaminate the sample during transfer and presentation to the instrument from, e.g., sample cups and pipette tips. Whilst potential for significant loss or contamination may be reduced because of the higher concentration, the general principles of clean handling should still be applied to ensure the integrity of the sample. In addition to the potential for contamination, losses of analyte may occur. An example of a volatile analyte with which losses can occur unless the analytical system has been rigorously tested is mercury; between generating the gas phase and the measurement step, losses can occur through the permeability of plastics or adsorption onto the equipment used.

7.13.1 Evaporative Losses

Other processes can also lead to changes in the concentration of analyte solutions between the sample preparation and the presentation of the solution to the instrument. If the storage of concentrates or samples is long, evaporation may be important with plastics such as LDPE, which are relatively permeable to water vapour. Pre-concentration steps often yield small final volumes and thus small changes due to evaporation can greatly influence concentrations; the use of refrigerated storage and low permeability plastics can help to minimize this problem. A further instance where evaporation of water from samples can be a problem arises with the use of automated sampler systems for instruments such as graphite furnace atomic absorption spectrometers. If slow analytical cycles are used, and relatively small volumes are present in sample cups, significant losses can occur (Figure 7.11).

The problem can be largely circumvented by having a well fitting cover to the autosampler tray, and if possible having a small pool of water within the enclosed tray area to maintain a high water vapour pressure and thus inhibit evaporation. The situation is even worse for volatile organic solvents which are to be analysed for their elemental content. If the organic solvent is immiscible with, and denser than, water, and no analyte transfer between phases will occur, it is possible to use a layer of water to inhibit solvent evaporation.

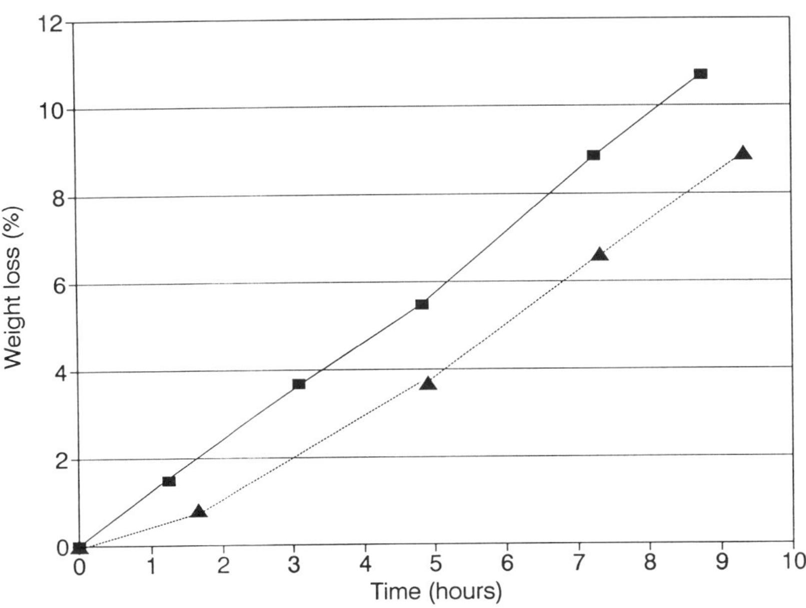

Figure 7.11 Evaporative weight losses from dilute acid solutions held in cups in an open autosampler tray. Initial sample volume was 500 μl. Mean relative humidities for squares and triangles were respectively 70% (21.4 °C) and 80% (21.7 °C)

7.13.2 Carry-over

A further problem which can be experienced is that of carry-over of analyte from one analysis to the next. A good example involves spray chambers in flame atomic absorption spectrometers; if a high concentration sample is followed by a blank or low concentration sample, the memory effect may lead to erroneously high values. Depending on the analyte concentrations present, and the geometry and flushing characteristics of the spray chamber, long flushing times may be needed to ensure that unbiased data are obtained. This may occur for example when Ca and Mg are to be measured at low concentrations in fresh waters if marine sediment digests with very high concentrations of these elements have been analysed by the instrument immediately before.

The most important point is the need to be aware of potential problems at all-stages in the analytical sequence, from the initial contact with the sample to the final reporting of the data.

7.14 QUALITY CONTROL

The objective of any trace analysis is to obtain accurate and precise data on the analyte of interest. The concepts of accuracy and precision are fundamental to quality control. *Accuracy* is a measure of how close the measurement is to the real value, whereas *precision* describes the spread of replicate or repetitive measurements. It should be remembered that a single measurement to assess accuracy will have some uncertainty associated with it because of the imprecision of the procedure; the more replicates that are measured to determine accuracy, the greater the significance that can be put on the estimate of the accuracy of the analysis. The systematic shift of a value from the true value is often termed *bias*. The assessment of the accuracy and precision of an analysis has to be an integral part of any good analytical programme. This is where a *quality control* programme, often working under a laboratory Quality Assurance management system, has a major role to play.

> Quality control must be an integral part of an analytical programme

The situation for implementation of quality control measures may vary from a research scientist dealing with a few samples of very low concentrations in unusual matrices to routine analyses at higher concentrations in larger analytical laboratories. The underlying rationale and approaches, however, apply in all cases, and extend from the reporting of the data back to the initial collection of the sample.

> Sampling and handling prior to analysis must be considered as a part of the overall data generation

Quality control is a major subject in its own right, and Keith *et al.* (1983) and Keith (1991) provide a good introduction and coverage. The focus here is on an overview of the quality control measures within the trace inorganic analytical laboratory. Aspects of quality control are essential to the effective solution of some analytical problems, and there are important links between this section and the following chapter on trouble shooting.

7.14.1 Implementing Quality Control Procedures

Errors in analysis can come under two basic headings; systematic errors, which are consistent in direction relative to the true value and lead to a bias in the data, and random errors, which may vary in size and magnitude but will tend to zero if enough measurements are made. The reduction of measurement errors to an acceptable and known level involves the consistent use of well trained personnel, adherence to carefully designed protocols, use of reliable equipment, use of calibrations and standards, and, where a team of analysts is at work, a designated person with an overview of the control of these factors. Many of these ideas come within the concept of good laboratory practice.

In any quality control programme there is a series of key analytical measurements to be made, as given in Table 7.8.

Which tests will be used and at what frequency will depend on the protocol used for the analysis, and this can vary significantly. Important factors are consistency of the matrix, the analyte concentration ranges, and the precision and detection limit criteria for the analysis. Guidance on protocols for quality control programmes can be found in Cheeseman and Wilson (1978), Taylor (1987), and Keith (1991).

7.14.2 Accuracy

It is now well recognized that, among any large group of laboratories attempting to carry out a particular analysis, at least some will obtain results which are a long way from the true result. In order to assist laboratories in the development of their methods, and to provide a degree of confidence in their results, reference materials have been developed. These materials are normally accompanied by a certificate, issued by a certifying body, describing the levels of certain known components of

Table 7.8 Essential measurements for a quality control programme (after Keith,1991)

Measurement	Purpose
Blank determinations[a]:	critical as detection limits are approached
'trip blanks'	test for contamination or losses prior to analysis
laboratory or method blank	sample matrix without analytes passed through *all* analytical steps
instrument blanks	sample minus analytes is passed through the instrument
calibration blank	to check if the matrix of the standard solution contains any analyte
Replicate measurements:	precision estimates
samples	estimates of precision of sampling and/or analysis
the same sample	to identify within laboratory precision only
the laboratory blank	for 'detection limits'; see below
Recoveries of analytes:	
spiked[b] laboratory blanks	check on contamination, calibration and interferences
samples	check for influence of sample matrix, such as interferences, and loss of analyte
Standardization:	accuracy checks
calibration standards	internal laboratory standards at comparable concentrations to samples used to monitor and calibrate the analytical procedure
laboratory control standards	Certified Reference Materials (CRMs)

a The use of blank experiments to track sources is discussed in Chapter 8, Trouble Shooting.
b Known additions of analytes are made to the sample.

the material. The certified levels are either determined by a single analytical laboratory employing a definitive method such as isotope dilution mass spectrometry, by a single laboratory employing two or more reliable independent methods of analysis, or by a number of independent laboratories coming to a consensus view over the analysis. The blind analysis of certified reference materials provides a valuable means of assessing the accuracy and precision of a newly developed analytical procedure, and can provide a suitable quality control sample for the routine checking of performance.

Reference materials are developed and supplied by a number of bodies including:

- Canada—National Research Council (NRC)
- China—National Research Centre for Certified Reference Materials (NRC-CRM)

- ESB-RM-F—Environmental Specimen Bank, KFA Jülich
- European Community—Community Bureau of Reference (BCR)
- France—Laboratoire National d'Essais (LNE)
- Austria—International Atomic Energy Authority (IAEA)
- Japan—National Institute for Environmental Studies (NIES)
- UK—Laboratory of the Government Chemist (LGC)
- USA—National Institute of Standards and Technology (NIST); previously NBS
- USA—The American Oil Chemists' Society (AOCS)

In selecting a suitable reference material, it is important that the sample matrix is as well matched to the samples as possible. A wide range of reference materials is now available for trace analysis, including the following.

Trace elements: blood (BCR); coal (BCR); fly ash (coal) (BCR, NRCCRM, NIST); foods (NIST, NRCCRM, BCR, NIES); glass (NIST); hair (NRCCRM, BCR); leaves (NIST, NRCCRM, BCR); sediments (BCR, NIST, NRCCRM, NIES); serum (BCR, NRCCRM, NIST); urine (NRCCRM); water (NRC, NIST).

Anions: Rainwater (+ cations + pH) (NIST); water (NIST, NRCCRM).

Speciation: marine biological (methylmercury) (NRC); marine biological (tributyltin, triphenyltin) (NIES); marine sediment (tributyltin, dibutyltin, monobutyltin) (NRC).

If no CRM exists for a particular application, then the quantitative recovery of spikes of the analyte from a sample can give an indication of accuracy, but if the spike which is added behaves differently from the analyte in the sample a good analysis is not guaranteed.

The use of CRMs will indicate only if the laboratory analysis is correct. If the analytes in samples have been lost, or gained through contamination before the analysis, the use of a CRM will not indicate accuracy. The use of careful blank experiments can help resolve the problem. Other approaches to inferring accuracy include, in the marine field, looking for generally smooth changes in concentrations and relationships of the trace elements with other well known parameters in the ocean water column ('oceanographic consistency'), and in work on ice, where consistent concentrations of elements in sequential concentric layers of cores is inferred to show that external contaminated layers have been removed.

Participation in inter-laboratory collaborative testing is a useful way to check that there is acceptable intercomparability between laboratories, but accuracy is *not* necessarily proven by good agreement. This approach is important if there are several laboratories making the same measurements on different samples within a large programme.

7.14.3 Limit of Detection

It is clearly important to know the lowest concentration of an analyte that can be measured, but much confusion can result from uncertainties in the way in which this value is obtained. The topic is involved and a variety of approaches and terminologies is used. These criteria are discussed by Keith (1991) and in Currie (1978), and only an overview and recommended procedures are given here.

The dilemma facing analysts in reporting results of analyses is to ensure that, whilst data with a low probability of being different from a blank value are not reported, potentially useful data with a significant probability of being different from a blank value are not discarded. The term in most common usage when discussing data from low level analyses is the *limit of detection* (LOD), which is the lowest concentration level that is statistically different from a blank at a specified level of confidence (Currie, 1988). Generally the criterion used is three standard deviations (3σ) of the blank value. The *method detection limit* (MDL) is also often used, particularly in North America; this is the minimum concentration of a substance that can be measured and reported with a 99% confidence that the analyte concentration is greater than zero, and is obtained by analysis of a sample in a given matrix containing the analyte (US EPA, 1984). The LOD and the MDL are thus quite different, as the MDL includes components from blanks and handling procedures; the LOD is equal to the MDL only when the blank is zero. Other approaches exist.

It is clear that differing practices can be used for differing applications and between laboratories unless clear guidelines have been followed. Because of these uncertainties, the best general approach is to present a full data set and to define the criteria used. Thus, if a LOD of 3σ of the blank value is chosen, it is important to quote the blank value, the standard deviation (and number of observations), and that the 3σ criterion has been used. If all sample data are presented, and those below the LOD are flagged, then the user of the data has the opportunity to apply other criteria if appropriate to his/her needs. If the MDL is used, it is still important to report the blank values for the method.

Deciding on criteria for the identification of 'outliers' in a data set can also be difficult. Generally these are analytical results so far from the average that there is a low probability that they are correct. Statistical tests can be applied to aid in their detection (Barnett and Lewis, 1978), and sometimes it may be evident from problems with sampling or procedures in the laboratory that the data are suspect (see Chapter 8, Trouble Shooting). When examining new environments or batches of materials, however, these apparently anomalous data may be correct and thus be highly significant in providing information on the system under study. Wrongly discarded data delayed the discovery of the hole in the ozone layer for some years. As with detection limits, it is important to include all data, for the outliers to be clearly flagged, and for the criteria used for their selection to be stated.

REFERENCES

Bachmann, K.(1981) *Crit. Rev. Anal. Chem.* **12**, 1.

Barnett, V. and Lewis, T. (1978) *Outliers in Statistical Data.* Wiley, Chichester, UK.

Cheeseman, R.V. and Wilson, A.L. (1978) Manual on analytical quality control for the water industry, *Technical Report TR66*. Water Research Centre, UK.

Cresser, M.S. (1978) *Solvent Extraction in Flame Spectroscopic Analysis.* Butterworths, London, 200 pp.

Currie, L.A. (1978) Detection: overview of historical, societal and technical issues. In Detection in Analytical Chemistry, Importance, Theory and Practice, L.A. Currie (Ed.); *ACS Symposium Series No. 361*. American Chemical Society, Washington, DC.

Drabaek, I. and Iverfeldt, Å. (1992) Mercury. In: *Hazardous Metals in the Environment*, M. Stoeppler (Ed.). Elsevier, Amsterdam.

Kantipuly, C., Katragadda, S., Chow, A. and Gesser, H.D. (1990) Chelating polymers and related supports for separation and preconcentration of trace metals. *Talanta* **37**, 491–517.

Keith, L.H. (1991) *Environmental Sampling and Analysis: A Practical Guide.* Lewis Publishers, Chelsea, MI, 143 pp.

Keith, L.H., Crummett, W., Deegan, J. Jr., Libby, R.A., Taylor, J.K. and Wentler, G. (1983) Principles of environmental analysis. *Anal Chem.* **55**, 2210–2218.

Kingston, H.M. and Jassie, L.B.(1988) Eds. *Introduction to Microwave Sample Preparation: Theory and Practice.* American Chemical Society, Washington, DC.

Landsberg, J.P., McDonald, B. and Watt, F. (1992) Absence of aluminium in neuritic plaque cores in Alzheimer's disease. Nature 360, 65–68.

Martin, J.-M., Nirel, P. and Thomas, A.J. (1987) Sequential extraction techniques: Promises and problems. *Mar. Chem.* **22**, 313–341.

Miller, J.M. (1975) *Separation Methods in Chemical Analysis.* Wiley, New York, 309 pp.

Mizuike, A. (1983) *Enrichment Techniques for Inorganic Trace Analysis, Chemical Laboratory Practice.* Springer Verlag, Berlin, Heidelberg, New York.

Morley, N.H., Fay, C.W. and Statham, P.J.(1988) Design and use of a clean shipboard handling system for seawater samples. Advances in underwater technology, *Ocean Sci. and Offshore Eng.* **16**, 283–289.

Morrison, G.H. and Freiser, H. (1957) *Solvent Extractions in Analytical Chemistry.* Wiley, New York, 269 pp.

Riley, J.P., Robertson, D.E., Dutton, J.W.R., Mitchell, N.T. and Williams, P.J. leB.(1975) Analytical Chemistry of Sea water. In J.P.Riley and G. Skirrow (Eds.), *Chemical Oceanography*, 2nd edn, Vol. 3 Acdemic Press, London, pp.193–514.

Schüssler, U. and Kremling, K. (1993) A pumping system for underway sampling of dissolved and particulate trace elements in near-surface waters. *Deep-Sea Res.* **40**, 257–266.

Stary, J. (1964) *The Solvent Extraction of Metal Chelates.* Macmillan, New York.

Taylor (1987) Quality Assurance of Chemical Measurements. Lewis Publishers, Chelsea, MI.

Tessier, A., Campbell, P.G.C. and Bisson, M. (1979) *Anal. Chem.* **51**, 844–851.

Tschöpel, P. (1992) Sample treatment. In M.Stoeppler (Ed.), *Hazardous Metals in the Environment,* Elsevier, Amsterdam, pp. 73–95.

US EPA (1984) Federal Register, Part VIII, EPA Guidelines establishing test procedures for the analysis of pollutants under the clean water act: final rule and proposed rule, 40 CFR Part 136.

Van Loon, J.C. (1985) *Selected Methods of Trace Metal Analysis: Biological and Environmental Samples,* Wiley, Chichester, 357 pp.

Wolff, E.W. (1990) Evidence for historical changes in global metal pollution: snow and Ice samples. In R.W. Furness and P.S. Rainbow (Eds), *Heavy Metals in the Marine Environment*, CRC Press, pp. 205–217.

8 Trouble Shooting

Problems can occur in an analytical procedure at almost any stage. The extent to which these delay progress is often governed more by the approach taken in tracking down the source than by the steps which are required to correct it. The ability to think through these problems as they occur frequently distinguishes the experienced analyst from the rest. In this section the emphasis will be placed on the identification of problems and how to deal with them. Errors occur in all scientific measurements; it is only an error over and above that which is acceptable for the purpose of the analysis which indicates a problem which has to be overcome.

> Analytical errors are indicators of a problem, not problems in their own right

Most trace analyses go through a particularly troublesome period in their development when things seem constantly to be going wrong. If there is no more fundamental reason for these problems, a critical appraisal of the working practices of the analyst may reveal their source. No one likes to be looked over in such a manner, and a critical self appraisal may be all that is necessary to identify the problem. If the problems occur erratically, fastidious note taking, often including times and even the containers used in the analysis, may help to identify the source of the problem. The following examples may serve to illustrate the rather unpredictable nature of some of the problems that we have encountered.

Event 1 Periods of high frequency noise on an instrument output.
Cause 1: Electrically noisy energy controller on a heating mantle.
Cause 2: the operation of a radio-frequency generator in another laboratory.

Event 2 Complete loss of a mercury absorption signal from an atomic absorption spectrometer which had been moved to a 'new' laboratory. The machine worked for other elements.
Cause: the previous laboratory occupants, electrochemists, had been using mercury for many years. The atmosphere was severely contaminated with mercury vapour which attenuated the spectrometer light beam.

Event 3 A sudden change in instrument sensitivity in the middle of a batch of samples.
Cause: inspection of the times when the measurements were made revealed that the analyst had a coffee break, leaving the machine unattended during that time. During that break the settings had been changed.

Event 4 A machine that had a noisy output during the day but improved significantly at night.
Cause 1: the gas chromatograph (GC) was in the same room as a poorly vented atomic absorption instrument operating with an air—acetylene flame. The GC detector was responding to the fumes.
Cause 2: electrical noise on the mains originating from other laboratories which worked a shorter day.

Event 5 Occasional antimony contamination of samples.
Cause: a single contaminated pipette that could not be cleaned by normal methods.

It would not be true to imply that all these problems were rapidly solved. Their diagnosis in most cases, required rather unusual methods or lateral thinking. The noisy heating mantle problem, for example, was identified by associating the instrument recorder response with an audible noise from the mantle. Noting the time that each analysis is carried out is more common in automated systems than in manual methods, but trends or step changes can often be observed on a time base which would otherwise not be evident. A procedure which is irreproducible may be cured by carrying out measurements 'by the clock' with a fixed time between readings. To identify a consistently contaminated vessel or piece of apparatus, it may be necessary to number and record the identity of each piece as it is used in a procedure.

An analysis is normally carried out to answer a scientific question. In all situations a judgement must be made as to whether the analytical error is a significant problem within the context of that question. There are two main types of analysis: those which require a determination to be carried out within prescribed limits of precision and accuracy, and those which employ a standard procedure to be carried out to judge whether the result obtained from a sample lies within prescribed guidelines. Strictly, only those analyses which can be demonstrated to be capable of producing accurate results should be employed and estimates of associated errors should be given. Unfortunately such methods are not always available: results are often produced which are of use only for comparative purposes, using the same method and frequently within the same laboratory!

Many of the analyses carried out today are for compliance purposes, and set out to ascertain whether the level of an analyte is below (or sometimes above) a prescribed control limit. Provided that it can be shown that a sample is within the control limits, no further use is likely to be made of the data. For many such studies random errors of 20%, which would be disastrous in the determination of the elemental composition of a new chemical, could be deemed to be acceptable. It is sometimes argued that it is better to be able to screen 20 samples with poor precision than to be limited to the analysis of one sample with high precision. Indeed, it is a recognized fact that in some analyses a bias is accepted as being present. Under such circumstances control limit measurements can be acceptable provided that the limits are set with this bias in mind. However, a major problem arises with such an approach. The comparison of results between laboratories becomes difficult unless all participating laboratories use a common method and have samples which do not differ in such a way as to influence the accepted bias. Biased analyses can be used for screening provided that they are used solely to identify samples for which further, more rigorous investigations are necessary. Whether such a practice is acceptable is highly debatable, but its use is often dictated by the pressures of the real world. Biased control limit analyses are at their most dangerous if the results are later released for purposes (or to persons) which fail to take into account the known limitations of the measurements.

It must be remembered at all times that there is a cost, in finance, time and labour, involved in improving the quality of data. It is therefore important that care is taken to identify the required performance of a procedure so that unnecessary refinements are not carried out. The time taken to improve a method to give a relative standard deviation of better than 1% may be longer than is available for the whole task. That said, the reporting and discussion of errors must be used to communicate the confidence the analyst has in the results. Unqualified results will always be interpreted by the untrained as being both 100% accurate and significant to the last decimal place. It is better not to record a result than to have a knowingly incorrect result permanently on record and incorrectly cited.

> It may be more important to know the magnitudes of your errors than to have minimized them all

The first step in the refinement of an analytical technique is the identification of the types of problems which are involved. These can be divided into two main classes based upon the errors they cause:

(1) Systematic problems; these give rise to a constant, fixed absolute or relative error in the final analytical result;

(2) Random problems; these give rise to random variations in the analytical result

8.1 SYSTEMATIC PROBLEMS

A systematic problem gives a constant error, the magnitude of which is independent of the sample which is being studied. It is possible for these to be a constant absolute error (high by, e.g. 1 μg/l) or a proportional error (e.g. 10% high). Identification of whether a proportional or absolute error is involved is an important step in identifying the source of the problem.

The most commonly encountered example of a problem which causes a systematic error is contamination by the introduction of analyte into the sample. This may be caused by the addition of impure reagents or by the use of dirty apparatus. For the purposes of this section, a sample container will be considered as one of the reagents. Since the concentration of analyte introduced in this way is normally fixed within a batch of samples, this is an absolute systematic error and blank analyses will generally show up the problem.

In most analytical procedures many components may potentially contribute to the problem, and identifying one responsible reagent out of all those which have been used in the procedure can be far from straightforward. A systematic approach to diagnosing the cause of the problem is highly recommended.

8.1.1 The Blank Analysis

The diagnosis of a number of analytical problems can be carried out with blank analyses (see previous chapter). In order to deal with the many possible aspects which may contribute to an analytical bias, various different blank experiments are possible.

The *system or instrument blank* is employed to establish the background response of an analytical system when no sample is present. A common example of this approach is to identify ‘ghosting’ or memory effects in an analytical instrument. This frequently occurs when the analyte is incompletely purged from the instrument between samples.

The *solvent or calibration blank* is employed to identify and quantify the presence of impurities in the reagents which are employed to make up the standards for the method. If the reagents are contaminated a calibration line will not pass through zero but will run parallel to the line obtained using uncontaminated solutions (Figure 8.1).

> Zero concentration standards should be included in all calibration exercises

The *reagent or method blank* assesses contamination which is introduced during the sample preparation and analysis. It is normally carried out by performing these operations on either no sample or a dummy sample which is known to be free of the analyte.

The introduction of impurities into a sample is one of the most commonly encountered problems in inorganic trace analysis. It is therefore important in all

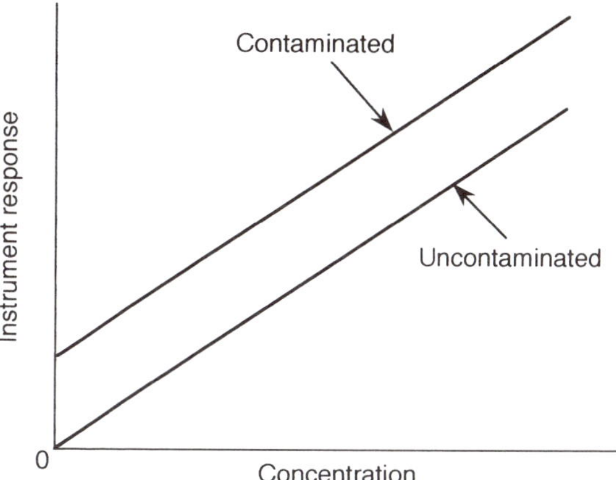

Figure 8.1 Calibration lines obtained using contaminated and uncontaminated standard solutions

batches of analyses to include a number of blanks. In its simplest form, the blank analysis can be described as the application of the complete analytical procedure to the analysis of no sample. In many cases this is straightforward. In the analysis of trace metals in paper, the digestion and analysis steps can be as readily carried out on no sample as it can on 1 gram of paper. Any metal which is present in the digest can then have arisen only from the digestion reagents, glassware, laboratory atmosphere etc. In this case, if the blank result is only a small proportion of the metal content of the paper, it is possible to correct the paper analyses for the blank contribution.

> The blank is as important as the sample.
> One is useless without the other

If no blank values are available all analyses are invalidated, as there can be no confidence in the accuracy of the results. There are circumstances, however, when a simple blank analysis is not possible. Consider the following example analysis.

> Take 1 l of seawater in a 2 litre separating funnel. Add 1 ml of buffered complexant solution followed by 10 ml of methyl isobutyl ketone (MIBK or 4-methyl-2-pentanone). Shake vigorously for 2 minutes, separate off the MIBK and analyse it for zinc by flame atomic absorption spectroscopy.

At first it may seem straightforward to carry out the blank analysis by substituting the seawater by high purity distilled water. However, this substitution does not necessarily achieve the desired result of appraising only those contributions from the reagent, glassware and procedure. If the high purity water is not sufficiently pure, the blank correction cannot be reliably applied—-there is no way of knowing whether the blank measurements are derived in part from the water, and it would certainly be very dangerous to assume that the water did not contain significant quantities of metal.

There is a second problem with this substitution. Seawater and distilled water have radically different chemical and physical properties. Seawater, for example, contains approximately 3% of salt by weight. This high ionic strength may significantly influence the formation of the metal complex and will alter the distribution coefficients that control the extraction of the metal complex into the MIBK.

How then can a blank analysis be carried out in such circumstances? It may in this case be possible to prepare 'zinc-free' seawater. Attempts to prepare artificial seawater from pure reagents are unlikely to be successful, as the levels of impurities in the reagents which have to be used will almost certainly be greater than those which are present in unpolluted seawater itself. If suitably uncontaminated seawater is unavailable, some means is required to purify the seawater without significantly altering the seawater matrix. The analytical procedure itself will often provide clues as to how this may be achieved. In our example the complexation/extraction procedure could be used to purify some seawater for the blank experiment.

> The analytical procedure may itself be suitable as a means of producing pure reagents

A little lateral thinking may sometimes be helpful in devising a route to the preparation of a dummy sample which is essentially free of the analyte. In the analysis of phosphate in seawater, a widely accepted method of preparing 'phosphate-free' seawater is to leave a container of unfiltered seawater in the light for some weeks to allow natural algae and bacteria to grow. The organisms consume the phosphate as a nutrient and can then be removed by filtration.

Having identified that the blank level is significant, the next task is to identify the aspect of the analysis which makes the major contribution. If any reagents are employed in the procedure, immediate suspicion must fall on them—if handling precautions have been taken, reagents will normally contribute more to the blank than storage and manipulation. There are certain groups of chemicals which are particularly prone to cause contamination:

(1) Reagents which are chemically related to the analyte or which are derived from a common source, e.g. selenium in sulphuric acid (both Group VI elements), arsenate in phosphate (both Group V), cadmium in zinc salts (cadmium is present as approximately a 1% component of zinc ores);

(2) Reagents having a high affinity for the analyte, e.g. chelating agents for metals;

(3) Reagents which are used in high concentration, e.g. acids used for digestions.

The task of refining procedures to minimize blank contributions should normally be directed at identifying and correcting contributions from the major factors only. Experience has shown that, whilst there is normally one major contaminating reagent or procedural step in an analysis, it is not unknown for several to contribute to the blank. This situation is however comparatively rare, and a search for a single major contamination source will normally either be adequate or will give rise to sufficient information to identify multiple contributions to the blank. To identify the sources of problems there are three main approaches. The reagents can be individually analysed or systematically substituted by reagents of known purity, or the concentrations of the reagents can be carefully varied.

Analysis of reagents

In many circumstances, the simplest approach to identifying those reagents which significantly contribute to the blank is to analyse the reagents individually. This may involve the use of a completely different analytical technique. Although this may not be appropriate to all the reagents, it will frequently cut down the number of chemicals which must be investigated by less straightforward means.

Substitution of purified reagents or reagents from different sources

A systematic substitution of reagents can be carried out if an alternative batch of reagent is available or if the reagents can be individually purified. This can be particularly useful in the diagnosis of blank problems which arise after a procedure has been in use for some time.

> Once batches of reagents have been found to be pure enough for an analysis, put some into store to help in the identification of future contamination sources

To assist in this approach it is recommended that, if a reagent is sufficiently stable, a reference batch of established high purity should be stored, to be used only for diagnostic purposes. It should not be assumed that obtaining alternative batches of reagent from different suppliers will guarantee that the material was manufactured by a different company or that it comes from a different production batch. Neither should it be assumed that a higher specification product will necessarily be any better than (or in fact different from) the cheaper product.

Experimental design can be helpful in identifying the rogue reagent. You may recognize the approach as being similar to a popular game involving the identifica-

tion of the positions and colours of hidden pegs.
Take the example of a procedure which employs four reagents: A, B, C and D. Carrying out only three blank experiments should identify a single major contaminant if the following substitutions are carried out.

Experiment 1: Use 'purified' A and original B, C and D.
Experiment 2: Use 'purified' B and original A, C and D.
Experiment 3: Use 'purified' C and original A, B and D.

If experiments 1 to 3 do not reveal the source of the blank problem, then D is the suspect reagent.

If the number of reagents is large, then more sophisticated treatments may be useful. Take for example the six reagent problem. If treated as above, five experiments might be necessary. An alternative approach would be to carry out the following experiments (Figure 8.2).

Experiment 1: In a single experiment use 'purified' samples of A, B and C and the original D, E and F. If this experiment results in a very significant drop in the blank then the responsible reagent is one or more of A, B or C. If not, it is D, E or F.
Experiment 2: If D, E or F is identified as responsible, use 'purified' D and original E and F. If the result from this experiment is low then the contaminated reagent is D. If the result is still high then the third experiment must be used to distinguish between E and F.

Experiment 3: Use 'purified' E and original D and F.
If experiment 3 gives a high result then the contaminated reagent is F; if low, then E requires purification or replacement.

Experiment	Original reagents	Purified reagents	Blank
	ABCDEF		High
1	ABC	DEF	Low
2	ABCEF	D	High
3	ABCDF	E	High

Reagent F is therefore impure

Figure 8.2 The identification of a single impure reagent from six by substitution

Experiment	Original reagents	Purified reagents	Blank
	ABCDEF		High
1	ABC	DEF	Medium
2	BCDE	AF	High
3	AFBD	CE	Medium
4	ABEF	CD	Low

Figure 8.3 The identification of two impure reagents from six by substitution

In this example a single contaminating reagent amongst six has been identified by three or fewer experiments. Unfortunately contamination does not necessarily arise from only one source. The approach is still the same but the number of experiments which is necessary and the level of interpretation may be greater. If the blunderbuss approach were to be taken of purifying everything then the blank would hopefully be improved, but the contaminated reagent would not have been identified. With little knowledge of where the problems lay, it would of course be necessary to purify all reagents in the future and it would not be possible to seek alternative sources of the offending reagent.

Now let us consider what happens when two impure reagents are present (Figure 8.3). The assumption will be made that the two impure reagents contribute approximately equally to the blank.

Experiment 1: Use purified D, E and F. If this results in a halving of the blank, impurities are being introduced by both groups of reagents (A, B, C and D, E, F).
Experiment 2: Use purified A and F. A high blank value here implies that both impure reagents are members of the group BCDE; A and F are ‘pure’. When experiment 1 is considered together with this result it becomes evident that (B or C) and (D or E) are impure.
Experiment 3: Purified C and E. A medium blank value leads to the conclusion that (A or B or D or F) and (C or E) are impure. There are only two solutions which satisfy both these conditions and the conclusion from experiment 2. The impure reagents are either (B and E) or (C and D).
Experiment 4: Purified C and D.A low result confirms C and D as the impure reagents. If this blank had been high then B and E would have been the culprits.

Increasing or reducing reagent quantities

If purified batches of the reagents are not available, a useful alternative approach to identifying major contributors to the blank is the systematic increase or reduction of reagent quantities. This approach depends on the reagents being tested in this manner being present in significant excess. It is important that changing the concentration of the reagent does not alter the analyte signal only by changing the method sensitivity. If this were to happen the investigator would wrongly conclude that the test reagent was significantly contributing to the blank. In order to minimize the possibility of such a problem, it is normally better to increase the concentration of the test reagent rather than to reduce it.

As far as the experimental design is concerned, systematically increasing or decreasing the reagent concentrations to identify the source of a reagent blank is identical to that described for the substitution experiments.

8.2 PROPORTIONAL PROBLEMS

Proportional problems typify equilibrium and kinetically slow reactions. They are associated with losses from solution or, for example, a signal enhancement by an interfering species which is present in the sample matrix.

8.2.1 Method of Standard Additions

Proportional problems are not shown up by carrying out blank analyses, as there is hopefully no analyte in the blank. In this case the method of standard additions may show up the problem, and in certain circumstances it can be employed to apply numerical corrections to the analytical result.

Let us take the example of the enhancement or depression of the signal in atomic absorption spectroscopy when this technique is employed for the analysis of a metal in a complex solution.

8.2.1.1 Single addition

In its simplest form, the standard addition experiment consists of increasing the concentration of analyte in the sample solution by a known amount (the standard addition). Provided that *no significant dilution of the sample occurs* when this addition is made, and *the interferent is in excess*, the difference in the measured metal content before and after the addition should equal the concentration increase expected from the quantity of analyte added. If not, an interferent is present and it is possible to calculate a better estimate of the correct concentration.

In the determination of magnesium in a soil extract using atomic absorption spectroscopy, an initial analysis carried out using external calibration with aqueous standards indicated that the sample contained 1 μg/ml of Mg. The concentration of magnesium in the solution was then increased by 0.4 μg/ml. This could be achieved by the addition of a small volume of a 100 μg/ml Mg solution to a known volume of the sample. Adding the extra metal in this way does not, within the errors of the analysis, result in a significant dilution of the sample matrix. Any interferent acting on the magnesium originally in the sample will therefore be expected to act similarly on the added magnesium, provided that the interferent is present in excess. In this particular example the expected Mg concentration in the new sample solution would be 1.4 μg/ml. If on re-analysis a result of 1.2 μg/ml were to be obtained it would be apparent that the interference effect was reducing the analytical result by 50% (a 0.2 μg/ml increase was measured when 0.4 μg/ml was added). A better estimate of the concentration of magnesium in the original sample would therefore be 2.0 μg/ml.

8.2.1.2 Multiple additions

Rather than this single addition, a series of additions is often made. This experiment can simply be thought of as preparing a series of standard solutions using the sample as the solvent. As before, it is important that adding the metal to the sample does not significantly dilute the sample, and the addition is therefore made by adding a small volume of a high concentration stock metal standard solution to the sample.

Consider a sample containing an unknown analyte concentration C, which gives an instrument response R_c. If the instrument response is linearly related to concentration and if in the absence of the analyte the instrument gives no response, we can write:

$$R_c = kC \text{ where } k \text{ is a constant} \qquad (1)$$

If we now add analyte to the sample to increase the concentration by x, a new response R_{c+x} is obtained and, as before:

$$R_{c+x} = k(C+x) \qquad (2)$$

Dividing (1) by (2) to eliminate k and rearranging gives

$$R_{c+x} = \frac{R_c}{C}x + R_c \qquad (3)$$

which is an equation of the form $y = ax + b$. When $R_{c+x} = 0$, x is equal to $-C$.
From a plot of R_{c+x} against the standard addition concentration increase (x), the intercept on the x axis will give the concentration of analyte in the original sample (Figure 8.4).
This is the method of multiple standard additions. It is simply a calibration of the instrument using the sample solution itself as a solvent for the preparation of the standard solutions. As the sample contains some analyte, the instrument response

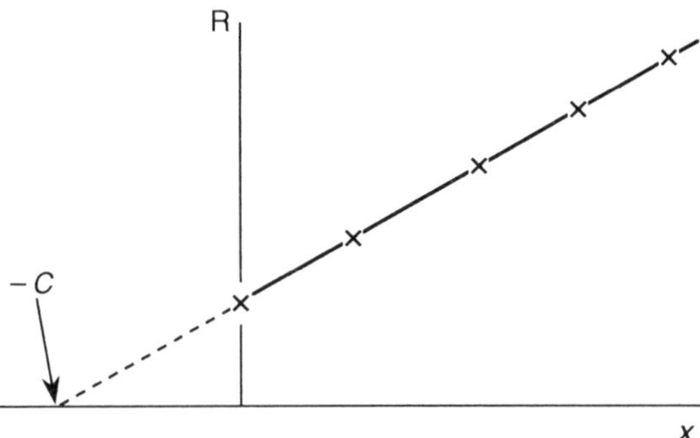

Figure 8.4 Extrapolation of the standard addition line to give an estimate of the analyte concentration in a sample

is increased, resulting in a vertical displacement of the calibration curve (Figure 8.5) by a constant amount.

The vertical displacement is being used to calculate the concentration of analyte in the sample. If there is no proportional interferent present in the sample solution, the standard addition line will be parallel to the calibration curve obtained using aqueous standards. As a result, the analyses by direct calibration and by standard additions will give identical results.

The presence of interfering species in the sample solution will lead to the gradients of these two lines differing (Figure 8.6), and it is under these circumstances that the method of standard additions will normally yield a more accurate estimate of the analyte concentration. There is a problem with the method of standard additions which means that results obtained in this way are subject to larger random errors than results obtained from interpolation using a conventional calibration curve. The method of standard additions involves extrapolation of the line towards

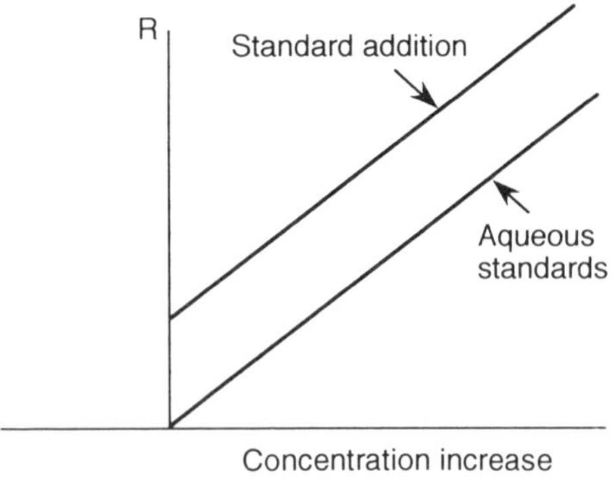

Figure 8.5 The standard addition line in the absence of interferences

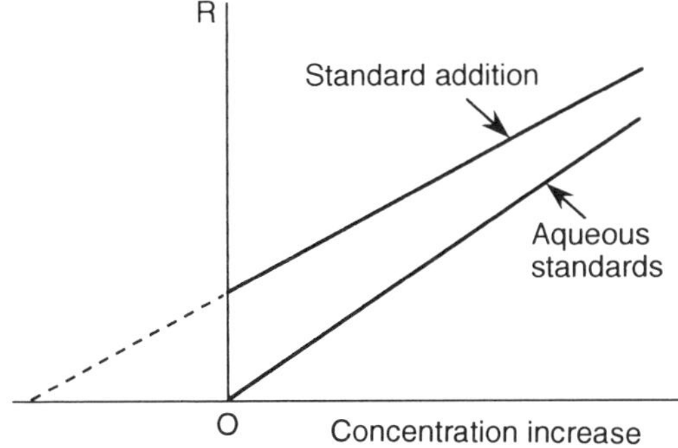

Figure 8.6 A standard addition experiment in which interfering species are present

the concentration axis, and the further this is away from the measured points the greater will be the leverage effect magnifying the experimental errors (Figure 8.7). It must be remembered that the approach taken above assumes that there is a linear relationship between the instrument response and concentration. This is not the case for all techniques. It is also appropriate only to those circumstances in which the interference results in a proportional enhancement or reduction effect. Any interferent which causes the enhancement or depression of the instrument response by a fixed amount (as in Figure 8.5) will not be detected by this method. The method will not, for example, deal with molecular absorption or scattering effects in flame atomic absorption spectroscopy.

8.3 RANDOM PROBLEMS

The most difficult problem to solve is normally the one which is irreproducible. Diagnosis is the process of looking for trends; apparently random events do not immediately appear to follow trends and do not therefore appear to be amenable to the same logic processes which are employed for the diagnosis of other difficulties.

In trace analysis the random problem can occur throughout the procedure. It may be that the reproducibility of the measurement is not of the desired quality. It may be that sometimes the whole experiment works and sometimes it does not. The occasional sample may produce an apparently erroneous result.

Apparently random problems can be broken down into a number of groups.

(a) Poor precision Re-analysis of the same sample solution produces different results.

(b) Intermittent batch failure. A whole batch of analyses goes wrong even when a well tried procedure is used.

(c) Rogue results. The inexplicably high or low analysis amongst an otherwise unexceptional batch of analyses.

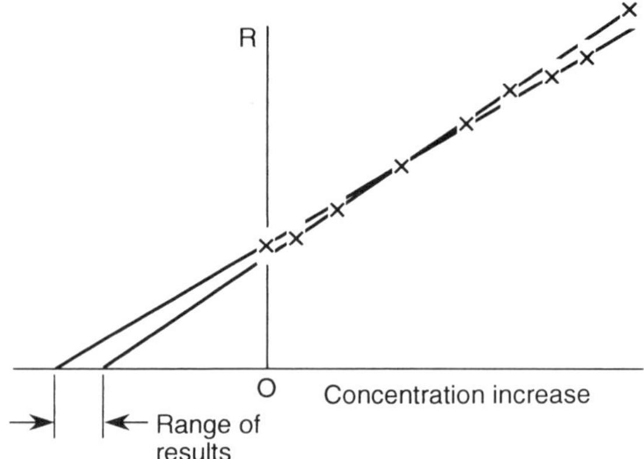

Figure 8.7 The leaverage effect governing the uncertainty of an analytical result

The refinement of an analytical procedure is at least in part a taming of the random elements by the identification of their causes. In any analytical procedure the random errors can be introduced in a number of steps:

- sampling and sub-sampling;
- sample handling and preparation;
- chemical pretreatments;
- the analytical finish (including instrument noise).

At each one of these stages analytical procedures have been developed to minimize both random and systematic problems which may arise. In this section it will be assumed that possibly the most important random element, sample selection and sampling, has been adequately dealt with in the experimental design. Problems which arise in the remainder of the procedure are better diagnosed by the replicate analysis of a single homogeneous sample.

8.3.1. Poor Precision

The concept of poor precision can only be identified from experience or expectation. It is therefore assumed that the experimenter has experience of what should be possible or wishes to achieve a specific goal. It must always be remembered that, whereas in some analytical procedures results may be reproducible to within 0.1%, to achieve reproducibilities within 20% may be a major success in more difficult analyses. It is important to distinguish here between the variability of sampling and the poor precision which results from a failure of the analytical procedure to generate the same result when presented with the same sample.

A number of possible sources of random error are likely to contribute to the precision of an analytical procedure.

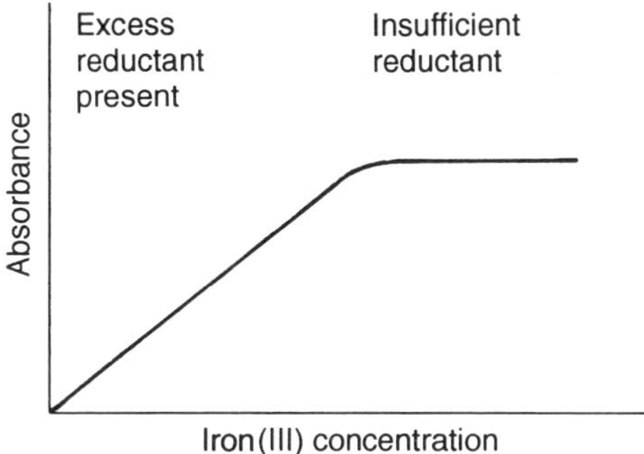

Figure 8.8 Non-linearity of a colorimetric procedure due to a critical shortage of reagent

Reaction based

In many chemical reactions, the concentrations of reagents are selected to be in excess. Take the example of the colorimetric determination of iron as its iron(II)—phenanthroline complex. As iron is frequently present at least in part as iron(III), this must first be reduced using ascorbic acid. If insufficient reductant is present the absorbance of the colorimetric reaction will not be a linear function of the iron(III) concentration (Figure 8.8).

In a robust analytical procedure an excess of ascorbic acid would be employed to ensure that the critical deficiency of reductant was never achieved. If this were not the case, a slight variation in the proportion of iron(III) in our example could lead to a wide variation in the analytical result. As the oxidation of iron is readily carried out by oxygen in the air or in solution, a random result could result from variable exposure to the air!

Instrument based

All instruments are subject to some degree of noise. This may be from the components of the instrument itself, such as a photomultiplier, or from the environment in which they operate. Although some of the instrument-derived noise is intrinsic to the design of the instrument supplied by the manufacturer, some component deterioration is to be expected and routine maintenance will be necessary. (Anyone who has owned a car will be aware that the driver is often the last person to be aware that the steering has deteriorated.) In a routine laboratory environment it is easy to monitor the performance of an instrument using quality control standards. In addition to the routine measurement of instrument sensitivity, regular replicate analysis of a single quality control standard allows the user to monitor the instrument reproducibility as it changes through its life; see, for instance, Figure 8.9.

The reproducibility of results which can be obtained using a specific instrument therefore becomes known, a new operator knows what to expect from the machine

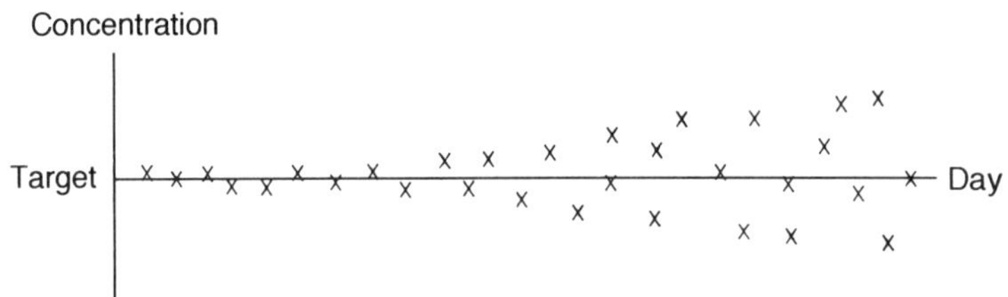

Figure 8.9 An instrument quality control chart showing deteriorating precision

and maintenance can be initiated when necessary. Such an approach can be used to predict imminent breakdown, so that suitable arrangements can be made. It is worth noting that the deterioration of instrument performance may not necessarily be gradual; this is especially true when the instrument has just been moved—the movement may have disturbed the optical alignment or dislodged an electrical connection.

To solve a problem of poor reproducibility it is necessary to break the analysis down into its component parts and to remove those elements which are the most probable contributors. The two ends of an analytical procedure, sampling/storage and the final determination, are often the best places to start. Replicate analysis of a simulated sample should identify the sampling/storage problem. Replicate analysis of a suitable instrument calibration standard will frequently reveal a poorly performing detection system (e.g. a noisy photomultiplier). If neither of these appears to be the source of the problem it is unfortunately necessary to carry out a systematic trace through the procedure.

8.3.2 Intermittent Batch Failure

One of the most frustrating experiences is the sudden failure of a whole batch of analyses when using a well tried procedure. Here it is necessary to look closely for deviations from the procedure, for reagent failure or unusual sample composition. Assuming that instrument failure has been ruled out, the next step under such circumstances is to identify all features of the analysis which have changed since it was last carried out. Good laboratory notes are invaluable in this exercise. Have any solutions been freshly prepared (or should they have been?). If so, was a new bottle of reagent used? Has a major change in apparatus been carried out, such as replacing glass bottles by polycarbonate equivalents? When all the more obvious explanations have been eliminated, look for the less obvious such as the use of an incorrect reagent—perhaps a nitrite rather than a nitrate. Having experienced a supply of sodium sulphite which was mis-labelled as sodium sulphate (and as a consequence destroyed a muffle furnace), we would stress that not even the laboratory suppliers can be assumed to be infallible.

8.3.3 Rogue Results

All analysts have at some time in their careers encountered the inexplicably high or low results amongst an otherwise unexceptional batch of analyses. The concept of duplicate analyses is there to deal with this eventuality, but if a pair of duplicate analyses differs there is of course no way of determining which is correct until further analyses are carried out. In the absence of the rare specific chemical interference which can cause this problem, its origin is often to be found in factors such as a faulty container or a speck of dust falling into the solution.

Index